CASS LIBRARY OF SCIENCE CLASSICS
No. 13

General Editor: Dr. L. L. LAUDAN, University of Pittsburgh

T0352721

A DISSERTATION

ON

ELECTIVE ATTRACTIONS

A DISSERTATION

ON

ELECTIVE ATTRACTIONS

TORBERN BERGMAN

SECOND EDITION

WITH A NEW INTRODUCTION BY

A. M. DUNCAN

Routledge
Taylor & Francis Group

LONDON AND NEW YORK

First Published 1785 by

FRANK CASS AND COMPANY LIMITED

Published 2014 by Routledge
2 Park Square, Milton Park, Abingdon, Oxfordshire OX14 4RN
711 Third Avenue, New York, NY 10017

Routledge is an imprint of the Taylor and Francis Group, an informa business

First Latin edition	1775
First English edition	1785
First French edition	1788
Second English edition	1970

ISBN 978-0-7146-1592-9 (hbk)
ISBN 978-0-415-76060-7 (pbk)

PUBLISHER'S NOTE TO THE 1970 EDITION

This is an exact facsimile reproduction of the 1785 English edition of Torbern Bergman's *Dissertation on Elective Attractions*. A new introduction by A. M. Duncan, and an index, prepared by the General Editor, have been added.

INTRODUCTION TO THE 1970 EDITION

BERGMAN'S *Dissertation on Elective Attractions* is itself an introduction to the tables of elective attractions which are its most famous feature, and to the author's own detailed exposition of the tables extensive notes were added by the translator, who was almost certainly Thomas Beddoes (see Appendix I). There is therefore no need for a modern commentator to add a further set of notes on each step in Bergman's argument. The need is rather to supply the context and sources of his thought without which it is difficult for the modern reader to understand its significance.

1. Bergman's Life and Times.

The Swedish military and political domination of the Baltic region ended with the death of Charles XII in 1718. Russia became the leading power in the region, though she was by no means unchallenged. In Sweden the Riksdag (Parliament) took all the legislative and most of the executive power from the crown and the Riksdag came to be controlled by a secret committee drawn from the three higher of its four estates. Although it was chiefly a country of small independent farmers, Sweden was in effect governed by the nobles at this time. France and Russia were rivals for political influence over Sweden, and Swedish life was in any case closely linked with Germany, so that culturally Stockholm was in touch with the main streams of European development. In Swedish politics there was a struggle between two parties—the Hats and the Caps. This period ended in 1771 with the accession of Gustavus III, who staged a bloodless revolution with the help of the three lower estates and assumed power from the aristocrats. He was assassinated by an army officer in 1792. Bergman's career was at its height in the earlier part of the reign of Gustavus III.

In spite of the political troubles, this was a period of great growth for Swedish industry, and a great age for Swedish literature and learning. Among its distinguished figures were the engineer Christopher Polhem, Linnaeus, Celsius and Swedenborg. In eighteenth-century Europe the arts generally depended for support and encouragement on kings and nobles. Gustavus was an especially active patron, and founded the Royal Opera, the Royal Theatre, the Academy of Literature, the Academy of Science, and the Academy of Art.

Bergman was born at Catherineberg in West Gothland in 1735. His father was a tax-collector, and sent him to the University of Uppsala near Stockholm to study law or theology. However, he strained his health trying to combine the study of mathematics and physics with enough law to satisfy the relative with whom he was living. For a time he had to leave the university. His first few papers were biological, under the influence of Linnaeus, but his dissertation for his master's degree in 1758 was on a mathematical problem in astronomy. He was made *magister docens* (assistant lecturer) in natural philosophy in 1758 and adjunct professor in mathematics and physics in 1761. While he held this post he published papers on the rainbow, the aurora borealis, and electricity, and a treatise on physical geography.

In 1767 Wallerius retired from the chair of chemistry, intending that a relative of his should succeed him. However, Bergman became a candidate for the post. He had a long-standing interest in chemistry, though he had published nothing in that field previously. He therefore published a dissertation on the manufacture of alum, which contained descriptions of quantitative experiments. Wallerius' friends violently attacked this work; but careful enquiries were made on the recommendation of the Crown Prince, later Gustavus III, and Bergman was appointed. As well as achieving a high reputation for research and teaching he served for a period as Rector of the University. In 1776 Frederick the Great of Prussia invited him to come to the Berlin Academy of Sciences at a much larger salary with fewer duties, but Bergman refused out of loyalty to Gustavus III of Sweden.

Bergman's health had been poor since he had strained it by overwork as a student, and very poor since 1769. He retired from active work in 1780 and died at the age of forty-nine in 1784. In such a short and busy life his output of scientific writing is astonishing. He married in 1771 the widowed daughter of a clergyman. She survived him, though their two sons died in infancy.

Probably the greatest advance made by eighteenth-century chemists was in basic knowledge of the variety and properties of inorganic substances. Like most chemists of his time Bergman was concerned with the practical application of his discoveries in medicine, mining and manufacture. His most valuable contribution was in methods of analysis. He gave the first detailed account of analysis by the blowpipe, and developed the first complete method of analysis of minerals in the wet way. He showed that metals could both be detected in qualitative analysis and determined in quantitative analysis by precipitation in a known compound without reducing them to metallic form. He analysed qualitatively and quantitatively a remarkably large number of minerals and mineral waters, although his own analyses were not very accurate and must be regarded as pioneer work. In particular he was a pioneer in classifying minerals by their chemical constitution rather than in the older way by their hardness, colour and crystalline form. He also investigated the properties of various gases. Throughout his career he was interested in the reform of chemical names and symbols, which will be discussed below as it arises in his work on chemical attraction. He was one of the comparatively few distinguished chemists who have written on the history of chemistry. Among Bergman's achievements should be included the discovery of Scheele, a poor apothecary whose chemical research was in some ways better than his own.

The original version of Bergman's *Dissertation on Elective Attractions* was published in the issue of the *Acta* of the Royal Society of Uppsala for 1775; but the translation here reprinted is that of a revised version which Bergman published in 1783. It was also translated into many other languages.[1]

2. Chemistry before the Time of Bergman.

Chemistry at the time of the publication of Bergman's *Dissertation* was at one of the most important turning points in its history. After a period of rapid growth it was about to be transformed by the great theoretical changes made by Lavoisier and Dalton into something that is recognisable as early modern chemistry; but it still retained some archaic features. There was no general agreement on the fundamental elements or principles which composed all other substances. Traces survived of the Aristotelean theory of the elements (originated by Empedocles) according to which all matter is of one substance and by taking various Forms can present itself to the senses through various properties. The four fundamental elements or varieties of matter are earth, water, air and fire, which enter into the composition of all other varieties of matter.

A later development of the Greek theory of the four elements was the theory of the three principles, salt, sulphur and mercury. It was introduced into European chemistry by Paracelsus, but had its roots in medieval Arab chemistry. The principles were not the ordinary salt, sulphur and mercury found in everyday use, but rather philosophical abstractions. According to the theory, all substances could be decomposed into these three principles on distillation. To the three active principles were often added two passive principles, *caput mortuum*, the inert residue left behind after distillation, and phlegm or water. Salt was the principle which gave substances the properties of fixity and incombustibility, mercury the principle which gave them the properties of fusibility and volatility, and sulphur the principle which gave them inflammability and oiliness. In origin these principles were not thought of as definite substances with properties of their own, producible in a specific quantity. However, the natural tendency of practical chemists was to simplify the concept and think of the principles as particular substances rather like elements in the modern sense. This made it possible for Robert Boyle and others to attack the whole theory, pointing out that the principles were not in fact producible from all substances, and did not explain the facts.

In its original form the theory of the three principles was a helpful model in explaining the facts then known; but by the eighteenth century its vagueness and inadequacy had made it necessary even for conservative chemists to modify it. Many used their own selection from the four Aristotelean elements, and the three active and two passive principles, sometimes adding acid and alkali. These were, however, usually included as salts. The general headings Earth and Salt become Earths and Salts, and among Earths were included the variety of minerals, and among Salts all the acids, alkalis and salts with which chemists had to deal in practice.

It was still assumed, nevertheless, that the elements or principles to be found in a compound substance could be discovered by analysing its properties, and that the properties of the elements or principles were reproduced with only slight modification in the compound substance. It followed that no gas could be thought of as forming part of a solid or liquid compound. Although Van Helmont in the first half of the seventeenth century had coined the word gas and shown that there were different kinds of gas, it was still usual in the early eighteenth century to assume that anything in gaseous form was air, possibly mixed with impurities which might give it particular properties.

Eighteenth-century natural philosophers tended to explain classes of natural phenomena by assuming the presence of some invisible, weightless, odourless and in fact otherwise propertyless fluid. Heat, light, and electricity, for instance, were generally believed to be such fluids. It was therefore natural to accept the theory of phlogiston, introduced by Stahl following the work of Becher. Phlogiston is not unlike the Aristotelean element fire, or the principle of sulphur. It explains why some substances burn and others do not. According to the theory of phlogiston, an inflammable substance contains phlogiston, which was sometimes called the 'matter of fire' or the 'principle of inflammability' When the substance burns, its phlogiston is given off into the atmosphere. Metals are supposed to be compounds of phlogiston and the calx, or as we should now call it the oxide, of the metal. When the metal was burnt or otherwise converted to its calx, the phlogiston was given off or combined with something else.

Scheme 55 in Table I in Bergman's *Dissertation on Elective Attractions* is a good example of a reaction interpreted in terms of the phlogiston theory. It is explained by the translator on page 380. The symbol on the left means nitre, or potassium nitrate, and the two symbols enclosed by the bracket to the right of it mean its two constituents. In eighteenth-century terminology these are nitrous acid and pure vegetable alkali, which in modern terms are nitric acid (not then differentiated from nitrous acid) and potassium hydroxide. The symbol on the right hand side of the scheme means marine acid, the modern hydrochloric acid, and according to Bergman its constituents are (in eighteenth-century terminology) phlogiston and dephlogisticated marine acid, now known as chlorine. The symbol in the middle means heat, and the symbol at the top of the scheme means phlogisticated nitrous acid, or, in modern terms, oxides of nitrogen. The whole scheme means that when nitre and hydrochloric acid are heated together, the nitric acid from the nitre and the phlogiston from the hydrochloric or marine acid attract each other more strongly than they are attracted by the potassium oxide or the chlorine. They therefore combine together to form oxides of nitrogen, which are believed to be a compound of the nitric acid with phlogiston.

This interpretation of the reaction was a useful one, for it explained the facts as far as they were known, provided a theoretical framework within which further hypotheses could be constructed and tested, and enabled the chemist to think of the reaction in terms of more or less distinct and identifiable chemical substances remaining unchanged through the reaction and forming new compounds by re-arranging themselves. The trouble is that phlogiston was not quite precise and definite enough, even in the hands of a sensible and practical chemist like Bergman. It was well enough known that many substances gain weight on combustion, which would seem to make untenable the theory that combustion consists of giving off phlogiston. The theory continued to be held because chemists were not yet accustomed to thinking in terms of weight, and even more because phlogiston was not necessarily seen as a definite substance having a determinate weight, or any weight at all. Not all chemists regarded as necessary the variants of the phlogiston theory according to which it either had negative weight, so that a substance became heavier when it was given off, or was so much lighter than air that a substance containing it was more buoyant and appeared to weigh less.

Another very different feature of eighteenth-century chemical theory was derived from the mechanistic outlook of eighteenth-century physics. Early in the seventeenth-century Gassendi had revived the ancient Greek theory that all the phenomena of the universe are to be explained simply in terms of the size, shape, motion and arrangement of atoms, which were surrounded by empty space. Other seventeenth-century philosophers, such as Descartes, had constructed similarly mechanistic explanations of the universe but, not wishing to commit themselves to the details of the atomic theory, they had called the invisibly small pieces of which they believed matter to be composed either 'particles' or 'corpuscles' instead of atoms. Boyle's was probably the most determined attempt to interpret chemical phenomena by a corpuscular theory; but the most influential work of this kind was Newton's, expressed particularly in the 31st Query at the end of his *Opticks*.[2] He there suggests the possibility that all chemical properties and reactions may be explained by the attractive powers of the particles of which matter is assumed to consist, as well as by their size, arrangement, polarity, hardness and other characteristics. Just as the attractions of gravity, magnetism and electricity operate over distances which are large enough to be appreciable, there may be other forces of attraction operating over insensibly small distances. A substance may have a stronger attraction for some substances than for others. Newton considers the possibility that at even smaller distances there may be a repulsive force between particles, which would account for the elasticity of vapours and gases. He rejects explanations based on the shapes of particles, such as Boyle's notion that the particles of air are like coiled springs and therefore tend to push each other apart, or the notion that particles combine together because they are hook-shaped.

These suggestions of Newton's were capable of being developed in many different ways, particularly because they are expressed as questions, beginning 'Are not . . .?' or 'May not . . .?' or 'Have not . . .?' Many different theories were put forward as following Newton's principles. So great was the prestige of Newtonian physics in the eighteenth century that it was generally felt that a proper science ought to be of the same kind—clearly based on geometrical or at least mathematical laws, whose consequences were explored mathematically and which depended on exact quantitative measurements. That was why theories involving the geometrical properties of atoms and the mathematical laws of force between them had such a strong appeal. However, they were put forward more often by mathematicians or physicists than by working chemists, for they conflicted with another fundamental assumption of eighteenth-century chemistry, that any theorizing must be soundly based on thorough, reliable experiments. Until the work of Dalton in the early nineteenth-century it seemed clear that about atoms, particles or corpuscles which were too small to be observed the chemist could have no experimental information, and could therefore make no statement which was not almost entirely speculative. Where chemists did permit themselves to speculate about atoms or particles, they were generally careful to make it clear that they were speculating and to keep the serious business of observed facts quite separate. Among the philosophers who did try to interpret the observed facts of chemistry in Newtonian terms, or what they claimed to be Newtonian terms, were 'sGravesande, Pemberton, Keill, Freind, Desaguliers, Knight, Lesage, Marzucchi and Boscovich.

An older tradition in chemistry of explaining why some substances combine and others do not was to speak either of their loving each other or having sympathy, or (particularly for the neutralization of acids by alkalis) of their being enemies and subduing each other. This attribution of human emotions to inanimate matter would be meaningless if it were metaphorical, and unjustifiable if it were meant literally. Boyle and other mechanical philosophers naturally objected to it. However, the word 'affinity' in the early eighteenth century still implies some family resemblance between substances which tend to combine. Only gradually does it lose this connotation. Chemists therefore had a choice between the word 'affinity', which might be criticized as attributing human qualities to matter, and the word 'attraction', which was criticized as attributing an active property, operating at a distance over an interval of empty space, to matter in which no evidence for such a property could be observed. French chemists in particular tended to avoid the word 'attraction' because in France Newtonian concepts did not replace the doctrines of Descartes until well into the eighteenth century. By the 1770s, however, the two words were virtually synonymous and meant simply 'tendency to combine' without commitment to any theory of the cause of this tendency. Bergman explains on page 7 his slight preference for 'attraction'.

Newton himself in his 31st Query had defined attraction in this empirical way, although at various times he did consider explaining it by means of the properties of an ether. He wrote:

> How these Attractions may be performed, I do not here consider. What I call Attraction may be performed by impulse, or by some other means unknown to me. I use that Word to signify in general any Force by which Bodies tend towards one another, whatsoever be the cause.

Nearly all chemists who define either affinity or attraction do so in such non-committal terms, and Bergman on the first page of the *Dissertation* uses words very like Newton's.

3. Chemistry in Bergman's Lifetime.

I. Chemical Theory.

The main theoretical developments in chemistry in Bergman's lifetime were progress in the use of quantitative methods, especially the habit of thinking in terms of the weights of particular substances taking part in a reaction, the evolution of a distinct and usable concept of the nature of a chemical element as opposed to a compound or a mixture, the establishment of the laws of constant composition and of the conservation of matter, and the elucidation of the nature of gases and of combustion. This last development naturally involved the refutation of the phlogiston theory. Most of these developments were brought to fruition by Lavoisier, but all were the result of a long series of contributors.

The use of weighing, the assumption that the weight of each reagent which entered into a reaction would be conserved throughout its course, and the assumption that the proportions by weight of the constituents in any given sample of a compound would be the same in any other sample of it, were natural in such technical processes as the assaying of ores, or the preparation of drugs by an apothecary. More academic chemists assumed if the question arose that the weights of each substance would be conserved through a reaction, but until the late eighteenth-century the question was not often considered. Principles and impalpable fluids would generally not be thought of as capable of being weighed even when more definite substances were. Isolated instances may be found before the middle of the eighteenth-century of chemical reasoning based on weighing; but Black's doctoral dissertation submitted at Edinburgh in 1754 is the first example of the systematic use of weighing as the basis of a new interpretation of a series of chemical changes.[3]

Black showed that when a mild alkali such as magnesium or calcium carbonate was heated it gave off a kind of air, and that it was the loss of this air and not the presence of any principle giving it caustic powers which allowed it to become caustic. The ratio of the weight of mild alkali to the weight of the caustic alkali which it yielded was always the same, and the caustic alkali when again combined with this special kind of air (i.e. in the case of quicklime by slaking and boiling with a solution of what is now called potassium carbonate) was converted back into the original weight of mild alkali. The air must therefore be combined with the quicklime or magnesia by chemical attraction, or fixed, and not just lodged between the pores of the solid as Stephen Hales had supposed. Black established the properties of the particular kind of air involved, which he called fixed air. Bergman called it 'aerial acid', and it is now called carbon dioxide. Thus it was shown for the first time that there were more than one distinct species of air, and that at least one of them could form part of a solid compound. Black also distinguished between temperature (sensible heat) and heat (specific heat), and discovered that different substances had different specific heats, and that there was latent heat in liquids and vapours. He measured the specific and latent heats of various substances.[4] However, he supposed, as was usual in the eighteenth-century, that heat was an elastic fluid which was held between the pores of substances which had absorbed it.

Henry Cavendish showed that what he called 'inflammable air' (later named hydrogen by Lavoisier) was another distinct kind of air and established its properties.[5] Since it is the calx of a metal which combines with an acid to form a salt, it was natural to identify the inflammable air which is given off with phlogiston, although not all chemists accepted so simple a view.

Oxygen (as Lavoisier later named it) was identified as another kind of air by Scheele, though Priestley discovered it independently and was the first to publish the discovery.[6] He called it 'dephlogisticated air', because it so readily allowed things to burn in it and so was presumed to be very ready to take up phlogiston. Scheele showed that ordinary air consisted of two 'elastic fluids' (as gases came to be called), one of them which he called 'fire air' being the same as Priestley's 'dephlogisticated air', and the other 'foul air'. The latter is not identified by Bergman as a distinct gas, though the experiments which the translator reports on pages 342–3 were on the verge of that discovery. Another name for 'dephlogisticated air' was 'vital air', because it was that part of the air which was good for breathing.

Cavendish showed that when hydrogen burnt in air it produced water, and Lavoisier showed that water was in fact a compound of hydrogen and oxygen. Thus it was demonstrated that neither air nor water was an element, and that there were a number of different kinds of elastic fluids or gases.

As is well known, it was Lavoisier who refuted the phlogiston theory by reinterpreting the gain of weight of certain substances on combustion, and who showed that combustion consists of combination with oxygen. He gave that name to the gas because non-metals when combined with it become acids (or rather, what are now called acidic oxides). Similarly, metals were now seen as elements which on combination with oxygen yielded bases. From the pattern thus perceived in inorganic chemistry Guyton de Morveau, Fourcroy, Lavoisier and Berthollet drew up their new system of nomenclature, which in essence is still used. In this system salts are given a binominal classification according to the acid and the base from which they are formed, and their names therefore show their composition, unlike the old names which either had nothing to do with the substance's composition or were misleading. Bergman's *Dissertation* was written before Lavoisier's new theory of combustion was generally accepted, but shows a stage in the development of inorganic nomenclature which immediately precedes the work of Lavoisier and his collaborators.

Bergman discusses heat and combustion on pages 229–278. He first considers the question whether fire, which he regards as the action of heat when increased to a certain level, is due to some kind of matter or to the motion of the particles of bodies. The latter view had been popular in the seventeenth-century; but Bergman in common with most of his contemporaries rejects it. He argues that motion is quickly halted, whereas a small spark will quickly increase to a large fire. There follows an account of the three chief current opinions on the nature of fire: (1) that fire is the same substance as light, which when it enters into chemical combination becomes phlogiston. This would explain the phenomena of combustion, at least at a superficial level; but Bergman objects that phlogiston has been shown to be the same as inflammable air. (2) that elementary fire is quite distinct from phlogiston, and the one expels the other from chemical combination. (3) that heat is a compound consisting of phlogiston and vital air. This third one is Scheele's theory, which Bergman tends to favour. It provides an explanation of the removal of free vital air from the atmosphere when something is burnt, and the contraction in volume of an enclosed body of air when something is burnt in it may then be explained by supposing that the heat formed passes through the glass of the enclosing vessel or is absorbed. Bergman later mentions an English theory that vital air and phlogiston combined to form aerial acid—that is, in modern terminology, that oxygen and phlogiston combined to form carbon dioxide. That would be a natural suggestion from the evolution of carbon dioxide from burning organic matter. This theory is associated with the Irish chemist Kirwan, whom Bergman apparently counts as English.

The reason that differences in specific heat of substances are not proportional to the differences in specific gravity is given as the variation in the attraction of the substances for heat and in the surface areas of their particles. Bergman expresses the variation of surface area with the number of particles in a given volume algebraically (pages 237–8). Melting and boiling occur when the atmosphere of heat surrounding the particles of a substance increases so as to separate them. This was a common notion in one form or another, the particles of heat often being thought of as repelling each other and so causing expansion.

Bergman refers to Lavoisier's demonstration that phosphorus and sulphur gain weight by combustion, and has no difficulty in explaining the fact, and even in suggesting a quantitative explanation, by supposing that heat as a compound of vital air and phlogiston is absorbed into them. It must be remembered that even Lavoisier regarded heat as a substance and supposed that it was emitted from a burning substance in the same way as phlogiston was. Charcoal is regarded as a compound of phlogiston and aerial acid (carbon dioxide), which would of course account for the production of heat and aerial acid when it is burnt. From the weight of aerial acid produced by a given weight of charcoal, after subtraction of the weight of the residue of ash, Bergman deduces the weight of phlogiston contained in it.

Clearly he is doing his best to treat phlogiston as an ordinary chemical substance, with all the normally expected properties such as weight and elective attractions. His thoroughness in subjecting the theory of phlogiston to calculations involving its weight is matched by his resourcefulness in reconciling the facts with the theory. The modern reader cannot help feeling that sooner or later he would have been faced with contradictory results which would have forced him to question the theory, or else to realise that phlogiston was allowed to have properties so elusive and indefinite that the theory was incapable of being falsified. Yet he is far from reaching that point. Indeed, the properties of phlogiston are hardly more vague than those of its great rival vital air and one or two of the other 'elastic fluids' at this stage of their history. We can hardly blame Lavoisier's predecessors for keeping their faith in phlogiston when they were also required to believe in the newcomers.

It was Lavoisier who first stated clearly the law of conservation of matter which had normally been an unstated assumption of working chemists for some time. Although Robert Boyle had given a very nearly modern definition of an element in 1661, it had not been generally noticed. Lavoisier's definition in 1789 of an 'element or principle' as 'the last point which analysis is capable of reaching' was similar, but gained general acceptance largely because he gave with it a list of substances which he believed to be elements. It is divided into four classes: the first consists of light, caloric (the matter of heat), oxygen, nitrogen and hydrogen; the second consists of oxidisable non-metals such as sulphur and phosphorus; the third consists of oxidisable metals such as antimony, silver and copper; and the fourth consists of the earths, chalk, magnesia, barytes, alumina and silica, which had not then been decomposed. It can be seen from the chapter headings of textbooks, and from tables such as Bergman's, that chemists had been groping away from the Aristotelean elements and the three principles towards some classification such as

Lavoisier's for many years. The law of constant composition, another unstated assumption of most chemists in the eighteenth-century, was first specifically stated by Proust at the very end of the century.

II. Affinity Tables.

To chemists who felt that their own branch of philosophy was undeveloped in comparison with astronomy or physics, and that a developed science ought to display order, mathematical reasoning precision, and enlightenment, chemistry before Lavoisier must have seemed untidy and unorganized, with no underlying pattern. Nowadays the periodic table is the framework of organization for inorganic chemistry. However, there was no generally acceptable theory of the nature of chemical combination or of the composition of matter to justify such a framework in the eighteenth-century, and any such theory which was assumed *a priori* would have been rejected as not supported by experiment. The correct procedure was to bring the mass of experimental results into order, and then to look for some general pattern which might be perceived in them. Tables of affinity were one way of trying to reduce such results into an ordered pattern.

The ancestor of all of them was that of Étienne-François Geoffroy, published in 1718.[7] It consists of sixteen columns in which the different substances are represented by the customary alchemical symbols. At the head of each column is the symbol for the substance or group of substances to which it refers. Below that the symbols for the substances with which it reacts are arranged in order of the strength of their tendency to combine with it, so that the nearest has the greatest affinity and cannot be displaced by any of the substances lower down but will displace any of them from combination with it.

Although the observation that some substances combine together more readily than others, with more or less detail about their preferences, was an old one, Geoffroy's table seems to be entirely original. No new table seems to have been produced before 1749 apart from Grosse's of 1730 which was merely a modification of Geoffroy's. Macquer's *Élémens de Chymie-Théorique* of 1749 reprints Geoffroy's table. Clausier in 1749 in his French translation of Quincy's *Pharmacopoeia* has a kind of table written in words instead of symbols, not set out in columns but printed continuously. A few new tables were printed in the

1750s and 1760s, almost all in columns of symbols like Geoffroy's, and a number appeared about 1775. Bergman's table (reprinted in the present volume) became the standard one. Erxleben's table of 1775, the same year as the first version of Bergman's, is in words and was followed by tables by Weigel and Wiegleb. There were few tables produced after 1790, and they were obsolete soon after the beginning of the next century. In general the tables tend to get steadily larger and more complicated throughout their history.

The idea of an affinity table is well expressed in the *Histoire de l' Académie Royale des Sciences* in commenting on the publication of Geoffroy's table in 1718:[7]

> That a body which is united to another, for instance a Solvent which has penetrated a Metal, should leave it to go and unite with another which is presented to it, is a thing of which the possibility would not have been guessed by the most subtle philosophers, and of which the explanation is still today not too easy for them. One imagines first that the second Metal fits the Solvent better than the former which has been abandoned by it, but what principle of action can one conceive in this better fit? It is here that sympathies and attractions would come very much to the point, if they meant anything. But in the end, leaving as unknown that which is unknown, and keeping to certain facts, all the experiments of Chemistry prove that a particular Body has more disposition to unite with one Body than another, and that this disposition has different degrees. . . . This Table becomes in some sort prophetic, for if substances are mixed together, it can foretell the effect and the result of the mixture, because one will see from their different relations what ought to be, so to speak, the issue of the combat. . . . If Physics could not reach the certainty of Mathematics, at least it cannot do better than imitate its order. A Chemical Table is by itself a spectacle agreeable to the Spirit, as would be a Table of Numbers ordered according to certain relations or certain properties.

In contemporary presentations of affinity tables the emphasis is on their usefulness in summarising chemical facts in an easily memorised form and enabling the user to predict the course of a reaction rather than on their theoretical implications. Geoffroy's table is indeed entitled *Table des Différents Rapports Observés entre Différentes Substances* (Table of the Different Relations Observed between Different Substances), evidently because *rapports* is a neutral word without the tinge of speculation about unverifiable causes which was carried both by the word 'affinity' and the word 'attraction'. However, as a summary of chemistry or even a classification of substances, it must be admitted that Geoffroy's and most later tables are a little disappointing. There is little evidence that they were ever of practical use as they were intended to be. Geoffroy's sixteen columns are headed Acid Spirits, Acid of Marine Salt, Nitrous Acid, Vitriolic Acid, Absorbent Earth, Fixed Alkaline Salt (i.e. sodium and potassium hydroxides), Volatile Alkaline Salt (i.e. ammonia), Metallic Substances, Mineral Sulphur, Mercury, Lead, Copper, Silver, Iron, Regulus of Antimony (i.e. metallic antimony), and water. These are far from being all the substances which were known, or even which were believed to be elementary.

The most natural and fundamental criticism of affinity tables as that they assumed a constant relationship between substances without evidence, or rather against a certain amount of evidence that affinities were not constant. In the light of modern knowledge one might say that there certainly is a constant relationship between elements and radicals, which has been expressed for instance in the electrochemical series; but this is far from being the only factor in determining the course of a reaction. The order of affinities shown by a single column in a table would therefore be insufficient to predict the course of all reactions between the substances included in it. Exceptions were soon pointed out—that is, reactions in which what was thought to be the normal order of affinities was reversed. The attempt to find general rules depended on finding a single order of affinities for a whole class of substances, as Geoffroy tried to do in the first column of his table which is for acids in general and in the eighth column for metallic substances in general; but the order for particular substances differed from the order which seemed to apply to their class in general, and so Geoffroy had to show separate columns for particular acids and particular metals.

Reversible reactions were also pointed out. Then it was realised that most reactions involved more than two substances, so that the result would depend on the sum or difference of competing affinities if it depended on such constant factors at all. The effects of heat and the effects of different concentrations of the substances concerned were also recognised as complicating factors. Attempts to allow for all these circumstances within the framework of a simple table were inevitably unsuccessful. In the *Mémoires de l'Académie Royale des Sciences* for 1720[8] it was already necessary for Geoffroy to defend his table against some exceptions which had been pointed out by his younger brother and by Neumann, chemist to the King of Prussia. He did so by suggesting that the substances involved were not quite the same as those included in his table, for instance that calx of lead is not the same as lead but a compound of it (contrary to the phlogiston theory of a slightly later period).

The table of C. E. Gellert in his *Anfangsgründe zur Metallurgischen Chymie*,[9] which is a handbook for the practical metallurgist with a few references to theory, is not an affinity table but a table of solubility, with the substances in each column so arranged that any substance would be precipitated from solution by the one below it, that is in reverse order of affinity. Gellert observes that the precipitation indicated by the table will not always take place in practice because the difference in solubility may be too slight, or because one of the substances which might be dissolved or precipitated dissolved the other, or for both these reasons together. At the bottom of seventeen of his twenty-eight columns he puts the symbols for some substances that are completely insoluble in the substances represented at the head of the column, and he shows some appreciation of the difference in solubility in 'the dry way', that is at high temperatures, and in 'the wet way', that is in aqueous solution at ordinary temperatures.

This point was taken up by the French chemist Baumé, who in his *Manuel de Chymie* suggested that there ought to be two separate tables, one for the wet way and one for the dry way, as the order of affinities was not the same in the two ways.[10] Bergmann was the first to follow this suggestion. The Scottish chemists Cullen and Black had also mentioned the effect of heat on affinities. Lavoisier, however, pointed out that there should really be a different affinity table for each degree of the thermometer.[11]

Anton Rüdiger in his textbook of 1756[12] remarked that he found the tables of Geoffroy and those who followed him deficient because they dealt with affinity as a thing in itself and not with the different circumstances that affected the combinations and precipitations attributed to affinity. However, this defect is not remedied in his own table except that he has a special section of ten columns showing substances that do not combine, rather like the insoluble substances which Gellert put at the bottom of his table. Similarly J. P. de Limbourg in his dissertation of 1761 on chemical affinities[13] considers compound affinities, i.e., cases of double decomposition where the affinities of more than two substances have to be considered. His table, however, shows only simple affinities in the normal way except that in four places he shows exceptions. Demachy and de Fourcy attacked the theories of affinity and attraction as unfounded, but both produced affinity tables of the usual type.[14] Demachy does have ten appendices to give special cases of the classes of substances dealt with in his main columns.

The only table of the period which is different from the others in its format is that of P. A. Marherr, who occupies most of his dissertation[15] in discussing the errors in Geoffroy's table. He claims that there are several genuine instances of doubtful or reciprocal affinity. Each of the 120 columns in his table contains only three substances, representing the fact that the second substance shown displaces the third from combination with the first. This arrangement has the disadvantage that much more space is required, as many repetitions are needed of the same symbol at the head of several columns to convey the same amount of information as Geoffroy's table conveys in a single column. The object is to be able to show more complexities, especially reciprocal affinities. These are shown by the many pairs of columns in which the second and third substances are the same but in reverse order, a rather ridiculous device. In fact the reader is left with the feeling that understanding of the factors at work in directing the course of reactions was quite insufficient to allow them to be represented in a systematic and explicit table.

In 1775, then, when Bergman produced the first version of his Table of Elective Attractions, the time was ripe either for a convincing demonstration that the whole idea of an affinity table was untenable or for a new table which would overcome all the objections to existing tables.

III. Double Affinity.

Bergman is particularly careful to point out that his main table deals with simple elective attractions, that is the attractions between pairs of substances, but that in most reactions three or more substances were involved. He therefore gives in Table I sixty-four diagrams showing the results of reactions in which there are three or four reagents. The earliest diagram of this kind occurs in the 1615 Paris edition of Jean Béguin's *Tyrocinium Chymicum* to explain the reaction of corrosive sublimate with sulphide of antimony[16]:

	Sublimated mercury	
Mercury		Vitriolic spirit
	Antimony	
Regulus		Sulphur

However, the notion that reactions may be the result of two or more competing forces, and attempts to represent such reactions by diagrams, are not found again till the eighteenth-century. They are characteristic of the Scottish school of chemists. Plummer, who was Professor of Chemistry at Edinburgh from 1726 to 1755, interprets some reactions as the result of competition between attractive and repulsive forces,[17] and Black in his *Experiments upon Magnesia Alba*[18] considers that interpretation of the double decomposition between magnesium carbonate and a calcium salt but prefers to believe in competition between different attractions because repulsion is doubtful. These interpretations owe much, obviously, to Geoffroy.

William Cullen, Plummer's successor at Edinburgh, used diagrams in which symbols for the substances taking part in a reaction were placed at the four corners of a square, or three of the corners, those which were combined before the reaction being joined by a bracket.[19] One or two arrows joined the substances between which the attractions were strongest, as read from Geoffroy's table, and showed the result of the reaction. This type of diagram closely resembles that used by Bergman except that Bergman uses no arrows or other symbols to show the forces acting, but merely shows which substances are combined at the start and finish of the reaction without any implications about the cause. Cullen also developed a second type of diagram, in which the substances involved were written at the ends of two lines (which he calls rods) forming a saltire. The attractions involved are represented by the letters W, X, Y and Z in the angles of the

Nitric acid   Silver

W
Y   Z
X

                                               Muriatic
        Soda   Acid

saltire. The affinity table would show which of the attractions were strongest, and so the course of the reaction. Black is said by Robison, who edited his Lectures on Chemistry after his death, to have used these diagrams (apparently after Cullen had introduced them),[20] but in one surviving manuscript of notes of his lectures actual numbers are given instead of Cullen's letters to represent the relative strength of the attractions involved. However, these numbers are arbitrary, chosen merely to make the sum turn out so as to predict the correct result for the reaction and not to represent the actual magnitude of any forces.

Later, however, evidently because he felt that such schemes implied mechanisms for which there was no evidence, Black used another kind of diagram. Two circles were drawn side by side, and a horizontal line bisected both of them. In the four semicircles thus formed were written the symbols for the four substances which took part in a double decomposition, or often for groups of similar substances, so that the diagrams summarised a number of similar reactions. In one diagram in which tin and silver, forming an alloy, are in one circle and iron and lead in the other, the second circle is divided by a double line to show that the iron and lead would not actually be combined. These diagrams are among the forerunners of the modern chemical equation. The French chemist Macquer in his famous *Dictionary of Chemistry* also has the suggestion that chemical combination is the result of superiority of one set of competing forces over another, but gives no diagram[21].

These attempts to represent chemical processes diagrammatically illustrate the general tendency of eighteenth-century philosophers, among whom the chemists tried not to be too much of an exception, to reduce the apparent disorder of nature to order and reason. Unlike the affinity tables, they were potentially fruitful attempts which are in the line of descent of successful nineteenth-century attempts to represent chemical processes and the composition of molecules by diagrams.[21]

IV. Quantitative Measures of Affinity.

Another natural line of thought in the eighteenth-century was to try to quantify chemical forces. The Dijon chemist Guyton de Morveau measured experimentally the forces required to separate plates of various metals from a mercury surface, which he thought would show the force of the attraction exerted by these metals in chemical reactions.[22] Achard, Professor at the Berlin Academy, similarly measured the force with which various solids adhered to various liquids.[23]

Another German chemist, C. F. Wenzel, measuring quantities which were at any rate chemical and not merely physical, constructed similar cylinders of various metals and allowed nitric acid to act on one face. From the weight of each which was dissolved in a given time he calculated the time which would be taken for the whole cylinder to be dissolved. He concluded that the affinity of substances for a common solvent was inversely proportional to the time they took to be dissolved, and so he calculated the relative strength of their affinity numerically. By analogy with mechanics Wenzel argued that the velocity of solution would depend on the nature of the metal, representing mass, and the affinity of the metal for the acid, representing force.[24]

Richard Kirwan of Dublin criticized Wenzel's figures as differing from the order of affinity known by other means. Kirwan based his own determinations of the relative strengths of the affinities of certain substances on the weights of base and acid required for mutual saturation. The work is important because it was a step towards the Law of Reciprocal Proportions, based no doubt on the work of Cavendish on equivalents. However, Kirwan confused the order of affinity of substances with the order of their equivalent weights, which was natural enough before Dalton's atomic theory had provided another explanation for differences in equivalent weights. Kirwan distinguished between affinities which tended to resist decomposition, which he called quiescent, and those which tended to produce decomposition and a new compound, which he called divellent, pointing out that decomposition would take place only if the sum of the quiescent affinities exceeded the sum of the divellent affinities.[25]

In the 1782 edition of his *Elements of Natural Philosophy* Elliott reprinted the 1775 version of Bergman's Table of Elective Attractions and added a column of figures intended to represent the actual magnitude of the affinities.[26] These were not found from direct experiments like those of Wenzel and Kirwan, but by assigning arbitrary numbers as Black had done for one reaction and adjusting them by trial and error to fit other reactions. However, they could not be made to fit all reactions and so the method failed. Fourcroy in 1784 produced a table of numbers representing the magnitudes of affinities by estimating the difficulty of splitting up a compound into its constituents.[27] However, this method failed for the same reason as Elliott's. Guyton de Morveau in the article 'Affinité' in his *Encyclopédie Méthodique* of 1789 reviewed all these attempts and made one of his own on the same basis as Elliott's.[28]

Except for the last two these numerical estimates of the magnitudes of affinities were published between Bergman's first version of his table in 1775 and the present version in 1783. He mentions Guyton de Morveau, Achard and Kirwan in his note on pages 4 and 5, but curiously enough not the rather obscure Wenzel, of whom he must have read in Kirwan's papers. In their empirical and quantitative study of reactions in solution they both had much in common with Bergman's thought. Although Wenzel was concerned with the velocity of reactions rather than with the proportions of the constituents in the resulting compounds, both his work and Kirwan's must have influenced Richter, the discoverer of the law of reciprocal proportions. Wenzel and Bergman, however, emphasize the importance of the concentration of a reagent for the velocity of a reaction. Berthollet later pursued this point still further in suggesting that substances might combine in any proportions whatever, according to their relative concentrations, and that the composition of compounds was not constant. This is a line of thought quite contrary to that pursued by Richter and Fischer; yet although Proust showed that Berthollet was wrong in rejecting the law of constant composition, it was the line which led eventually to the law of mass action.

V. Classifications of Affinities.

We need not here discuss in detail the various mechanical explanations of chemical attraction in the Cartesian tradition or along Newtonian lines which were suggested in the eighteenth-century, since Bergman hardly considers them. Chemists were on the whole inclined to think that chemical attraction was similar to gravitational attraction and connected with it, but not the same; but as the question did not seem to have much practical importance or to be capable of solution by any feasible experiments they tended not to spend much time on it. It was the mathematicians who tended to identify chemical and gravitational attraction completely.

Buffon in his *Histoire Naturelle*, possibly influenced by the rather similar ideas of Limbourg, tried to show how the inverse square law of gravitation might be modified when it applied to very small particles in contact.[29] The laws governing affinity, he wrote, are the same as the laws governing very large bodies. When the distances between the bodies are very large, their shape is of almost no importance; but when the distances are very small their shape is extremely important because the distances between the parts of a body are appreciable in comparison, and the total effect of the masses of the various parts of a body is not the same as if they were all concentrated at a point. Although in Buffon's time the shape of elementary particles was unknown, he predicted that future generations would be able to establish the law of attraction which effectively applied to particular substances and hence to calculate the shape of their particles. Macquer followed Buffon. Bergman echoes this argument and hence distinguishes between gravitational attraction on the large scale and contiguous attraction, which includes chemical attraction.

A distinction had long been made between cohesive attraction, or affinity of aggregation, which held together particles of the same substance, and chemical affinity or attraction which held together particles of different substances. Macquer gave the following classification of affinities in his *Dictionary of Chemistry*.[30] It might better be called a classification of circumstances in which affinity shows itself.

(1) Affinity of Aggregation, by which parts of the same kind are united.

(2) Affinity of Composition (i.e. chemical affinity), by which parts of different kinds are united to form a compound. These two types of affinity are both types of simple affinity: the remaining classes are types of Complex Affinity, which are all varieties of Affinity of Composition. Macquer describes them in words, but it is much more convenient for us to use letters as symbols, in the way which Bergman was among the earliest to use.

(3) When a principle A forms a compound of three principles with a compound BC because the affinities of each of the three principles for each other are equal or almost equal.

(4) When a principle A forms a new compound with a compound BC because A has an affinity for B nearly equal to B's for C. This is Affinity of Intermediums, B being the intermedium.

(5) If a supervening principle A has no affinity for B but a stronger affinity for C than B has, then the compound BC is decomposed and a new compound AC is formed.

(6) Sometimes when a new compound AC has been formed, B may again displace A and reform the original compound BC if A and C have almost equal affinity for B, the separation being brought about by particular

circumstances relating to some of their properties. This is Reciprocal Affinity.

(7) Double Affinity is involved when four principles are concerned in a reaction, and the result depends on the sum of the affinities between possible pairs of principles. It can happen, therefore, that two principles, neither of which singly could decompose a compound, do so when they act together.

Baumé, who was Macquer's assistant, gave a rather similar classification,[31] as did Guyton de Morveau,[32] and this had become the conventional pattern. Bergman's classification on pages 5 and 6 of the *Dissertation* is simpler but on the same lines, and Macquer's notion (derived from Marherr's) of particular circumstances causing Reciprocal Affinity resembles Bergman's much more developed list of the causes of apparent exceptions to the rules of affinity.

VI. Chemical Names and Symbols.

Before Bergman's time chemical nomenclature was quite unsystematic. The names of substances were either those in everyday use, such as soda or tartar, or referred to some property which had nothing to do with the chemical nature of the substance and might be misleading, such as Prussian Blue, Sugar of Lead, and Butter of Antimony. Bergman's work on chemical nomenclature has been thoroughly discussed by Dr. M. P. Crosland, whose book should be consulted for a fuller treatment than we have space for here.[33] It was well known in the eighteenth-century that neutral salts were formed by the combination of an acid with a base, and Rouelle had established the existence of acid and basic salts as well as truly neutral salts. However, some newly discovered salts had no names of their own, and those which had traditional names were still known by them.

Bergman had studied under Linnaeus, who had shown the possibility of a complete reform of biological nomenclature by giving each plant a name of two Latin words showing its genus and its species, thus substituting order and reason for the previous anarchy. He still allowed some of the traditional names where they were too well established to be forgotten or where they referred to some obvious characteristic, but did away with long descriptive names. Bergman was well aware, therefore, of the importance for scientific thought of systematic nomenclature, and suggested naming each salt by the acid and the base (which

might be an alkali, an earth or a metal in contemporary terminology) from which it was formed. In his paper *On the Aerial
Acid*[34] (the name which he gave to what Black had called 'fixed
air', now known as 'carbon dioxide') he suggested in 1773 calling
its salts by the name of the base with the adjective 'aerated' (in
Latin *aeratus*). Thus potassium carbonate would be 'aerated
vegetable alkali' and zinc carbonate 'aerated zinc'. In 1775 he
applied this principle to other acids and so produced such names
as 'vitriolated clay' (aluminium sulphate) and 'muriated lime'
(calcium chloride). However, he still uses some of the old names,
such as 'digestive salt' (potassium chloride) or '*plumbum corneum*'
(lead chloride). (A glossary of names found in the *Dissertation*
forms Appendix II to this Introduction). The names of the acids
and bases are not systematized, there being no clear theory of
their nature on which to base a system. Bergman went a little
further than this in his revised system of mineralogy in 1784, but
the nomenclature of the *Dissertation* remains essentially in the
1775 state. He influenced Guyton de Morveau, who collaborated
with Lavoisier, Fourcroy and Berthollet in a further reaching
reform of nomenclature which was nevertheless similar to
Bergman's in essence.[35]

The symbols which Bergman used in his tables are of a kind
which had a very long tradition, though they were much more
used in the eighteenth-century than in the previous century,
perhaps in imitation of Geoffroy. They were as well known and
as familiar to eighteenth-century chemists as the letters used as
symbols for the elements are to twentieth-century chemists.
The letters were introduced by another Swedish chemist,
Berzelius, in 1813, and soon made the alchemical symbols obsolete.
They were in any case an anachronism as soon as Lavoisier's
system of elements was established, and incapable of being
adapted to fit Dalton's atomic theory.

Bergman gives a key to the meaning of his symbols in the
first of the tables at the end of the book. The table is divided into
acids, alkalis, earths and metallic calces, with a section after the
earths for substances which did not fit into these classes, such as
water, phlogiston, matter of heat and oils. This classification is
not unlike Lavoisier's classification of the elements. The symbols
are of various kinds:

    (1)   Initial letters, such as No. 40 for spirit of wine (*Spiritus
           vini*) or No. 8 for acid of fluor.

    (2)   Very few of the signs are pictorial, and several signs

which are pictorial represent pieces of apparatus or processes and do not occur here. No. 42 for oil, which is very ancient, is no doubt a picture of drops of oil.

(3) The traditional alchemical signs for the metals, which are also the zodiacal signs for the planets with which they are associated. Such are the signs for gold (in No. 44), which is a symbol for the sun, and for iron (in No. 50), which is the symbol for Mars.

(4) Purely arbitrary signs, such as the circle with a horizontal line across it and a dot in each half to represent alkalis (nos. 26–28), and the symbols for vitriolic, nitrous and marine acids (Nos. 1, 3 and 5), all of which are probably derived from the traditional symbol for salt (a circle with a horizontal line across it). The symbols for newly discovered metals are usually modifications of the symbols for the traditional metals.

(5) Signs formed by combining two or more of the others. No. 30, the symbol for quicklime, is made by adding to the ordinary symbol for lime the letter p for pure. The symbol for lime, which in Latin is *calx*, is also used for calx in the other sense and added to the traditional symbols for the metals to mean their calces (Nos. 44–59). Bergman clearly does not mean by this that the calces are compounds. The symbol for alkali is added to the symbol for sulphur to mean liver of sulphur or 'saline hepar'. (No. 39). All the acids in the pure state are represented by a cross added to the usual symbol, which is used on its own for the combined state. Phlogisticated vitriolic and nitrous acids are represented by the usual symbol for the acid followed by the symbol for phlogiston, and dephlogisticated marine acid by the usual symbol for the acid followed by the symbol for phlogiston upside down. The symbol for phlogiston, which was used by Geoffroy before the phlogiston theory was current to mean 'oily principle or sulphur principle', is a combination of the symbols for sulphur and for oil (No. 36). In the tables Bergman sometimes uses a combination of the symbol for the acid and the symbol for the base to represent a salt (e.g. in column 34 of the table of Single Elective Attractions, lines 6, 7, and 15–18), in accordance with his ideas on nomenclature.

Certain symbols occur which are not in the key. The symbol for corrosive sublimate of mercury in line 19 of column 34 is a combination of the symbol for mercury with the usual symbol for sublimation. In Table I, the Table of Double Elective Attractions, occur the symbols for pure salt (Scheme 3), nitre (Scheme 22, though it is shown as a combination of vegetable alkali and nitrous acid in Scheme 10), cinnabar (Scheme 47) and sal ammoniac (Scheme 44). In Scheme 37 there is an elaborate compound symbol for 'tincture of vegetable alkali'.

In spite of the lack of system in them, these symbols are a surprisingly concise and clear way of representing reactions once the reader has learnt them, though some of the individual symbols are liable to be confused with each other. For instance, in line 13 of column 36 the symbol for antimony has a stroke missing and could easily be mistaken for the symbol for manganese.

4. The Dissertation on Elective Attractions.

After defining attraction and classifying different kinds of attraction, Bergman considers the fundamental question 'whether the order of attractions be constant'. He does not, of course, approve of 'those general rules which affirm, that earths and metals are in all cases precipitated by alkalis, and metals by earths, for they are often fallacious'. Nevertheless, he believes that for each particular substance the order of affinities really is constant, and that all apparent difficulties in accepting its constancy can be explained away. The order found in the wet way is the true order, and the differences which are found in the dry way when the temperature rises above a certain point are merely variants. Fundamentally, this is a matter of faith, over which Bergman exercises a free choice, and not a question of reason.

He next considers the mechanism by which heat produces changes in attraction. He attributes the change to differences in volatility of the substances involved, which is not altogether original. It had often been suggested, for instance, that the particles of a substance were surrounded by particles of heat, which repelled each other and so overcame the force of attraction. Bergman's account is different from most non-chemists' accounts of mechanisms inasmuch as he does not speculate about unobservable properties of particles nor write in merely general terms. He describes the way in which differences in volatility might operate by the most precise and powerful method available, that is by using algebraic symbols. Although various Newtonian writers such as Keill and Freind had done that before, it was still very rare for a chemist to do it. Bergman, however, uses the device freely and effectively.

Pages 18 to 63 of the *Dissertation* explain that the various suggested kinds of exception to the rule that the order of elective attractions is constant are only apparent. The first kind are 'apparent irregularities from a double attraction', where four substances are involved and so more than two simple attractions are competing. Bergman insists in particular that when a metal is precipitated from solution by another a double attraction is always involved, since the added metal and the precipitated metal are compounds of phlogiston.

There follow discussions of apparent exceptions from 'a successive change of substances', where one of the substances involved undergoes some other change before the reaction which seems to be anomalous, 'from solubility', where a newly released compound is not apparent because it is dissolved, 'from the combination of three substances', where one substance forms a new compound with a compound of two others instead of displacing one of them, and 'from a determinate excess of one of the ingredients'. It is here that Bergman recognizes that a large quantity of a substance may produce an effect which a small quantity of it could not do, and analyses how that might happen.

As we have seen, the various apparent exceptions to the constant order of attractions, and the ways of explaining them, are to be found in earlier writers on the subject. Bergman, however, is original in the thoroughness with which he develops the case and for the insight with which he interprets the phenomena observed in the laboratory.

He next describes the method by which he determined the single elective attractions and some of the additions which he had made to previous tables. It is here that he gives his famous calculation that even though it included only a few substances 'the slight sketch now proposed will require above 30,000 exact experiments before it can be brought to any degree of perfection'. However, as life is short and health is unstable, he has decided to publish it as it is, however defective. He died the year after the revised version of the *Dissertation* was published.

Pages 75 to 320 are occupied with a detailed explanation of the Table of Single Elective Attractions. The original version of 1775 had only 50 columns and was printed on a single sheet, but the revised version has 59 columns and needs two sheets. The added columns are those for acid of benzoin, acid of amber, acid of sugar of milk, acid of milk, acid of fat, acidum perlatum, acid of prussian blue, matter of heat (which has been discussed above), and siderite. As Bergman explains on pages 82 to 86, in his original version as in all previous tables the metals had been included as combining with acids; but as it is in fact the calx of the metal which combines with the acid in each case, the phlogiston being separated, it is the calx which should be placed in the table of single elective attractions.

Considerable light is thrown on Bergman's thinking about phlogiston by his *Chemical Dissertation on the Differing Quantity of Phlogiston in Metals*,[36] which is based on the principle that 'metals are precipitated from acids with the help of other metals by a double attraction'. The general reaction between two metals, an acid and phlogiston is discussed by using letters as symbols, and Bergman concludes that the quantities of phlogiston in the metals are inversely as the weights of each involved in the precipitation. From experimental determinations of the weights of various metals required to precipitate a given weight of silver in acid solution he can thus calculate the relative weights of phlogiston in the metals, and from another calculation (repeated on page 269 of the present *Dissertation*) he deduces the absolute weight of phlogiston in charcoal and hence the absolute weight of phlogiston in the various metals. Reading this paper, which was defended in 1780 between the appearances of the two versions of the present *Dissertation*, one feels more than ever that sooner or later Bergman would have discovered an inescapable inconsistency in the phlogiston theory.

The columns of the table of single elective attractions follow the same order as the key to the symbols, and might easily be divided into a section for acids, a section for earths, a section for substances such as water and phlogiston which do not fit into the other classes, and a section for metallic calces. A number of newly identified organic acids are included, though their composition was not then understood. Within the class of acids, occupying the first twenty-five columns, the order of attractions is very nearly the same, and there are strong similarities within the other classes. From the point of view of the modern chemist the similarities are natural; for we perceive the same orderly pattern of relationships underlying every column, and see the differences between columns as due to a variety of other factors.

Bergman, however, having no deeper foundation for his system than observation of the results of particular reactions, dare not assume a simple pattern and has to follow all preceding makers of tables in enlarging their scope to include more and more permutations of substances which behave a little differently from the general pattern for their classes. However, Bergman has laid out his column so as to emphasize the similarities between them.

Anyone who contemplated making an even better table after Bergman's must have felt a sinking of the heart: the task was endless, and indeed it is now clear that it was an impossible one. Weigel, Wiegleb, Gergens and Hochheimer, and Gren, the stouthearted Germans who attempted it, followed Bergman in having separate tables for the dry way and the wet way, but did not have any greater success. Weigel, Wiegleb and Gren also distinguished between single and double affinities.

Within the columns of Bergman's table a large proportion of the symbols are not separated by horizontal lines. This means that the substance at the head of the column combines equally easily with all of them, and so within these groups little information is given about elective attractions. The main innovation in Bergman's table, apart from its sheer size, is the division into an upper table for the moist way and a lower table for the dry way. Although in his text Bergman has written of the order of attractions being modified by heat in the dry way, in fact the substances included in each column for the dry way are by no means the same as those shown as combining in the wet way.

The translator has made some alterations in the table in transcribing it into English words, and since the modern reader may well rely on his version, it will be useful to record them here:

Col.   1: Bergman had placed 'pure ponderous earth' (barium oxide) in line 32; the translator has moved it to line 34 (see notes on pages 323–4).

Cols. 3–7: The translator has reversed lines 2 and 4, with question marks.

Col.   10: The translator has omitted siliceous earth from line 40.

Col.   23: The translator has added phlogiston in line 26.

Col.   25: The translator has omitted unctuous or expressed oil and essential oil from lines 26 and 27.

Col.   33, line 3 and Col. 38, line 9: The translator has stated 'pure vegetable alkali' although the symbol has no 'v' to indicate 'vegetable; but in

other places where there is no 'v' or 'm' the
translator has stated simply 'pure fixed alkali'.

Col.   33: The translator has omitted acid of borax from
line 47 and *acidum perlatum* from line 49.

Col.   44, line 12; col. 45, line 16; col. 46, line 18; col. 48,
line 17; col. 50, line 8; col. 53, lines 8 and 9: The
translator has correctly inserted the words 'acid
of' although in the table of symbols the cross
meaning 'acid' has been omitted.

There are also one or two minor variations in the names used.
The translator has written 'clay' in his transcription of the table
itself, but in the key to the symbols has literally translated
Bergman's phrase as 'argillaceous earth'. Although it is perhaps
in better style to call clay 'clay', Bergman has tried to be
systematic in calling all the earths 'earths' and avoiding common
names. The translator uses the phrase 'expressed oils' in the
table but 'unctuous oils' in the key to the symbols, and 'saline
liver of sulphur' in line 1 of column 39 (for instance) but 'saline
hepar' in the key and in line 7 of column 30. There is no difference
of meaning in either case. There are also one or two places where
the table of symbols in the English edition differs from the
transcription in words, for instance column 51, line 33, where the
symbol for copper is incorrectly given instead of that for mercury,
and in column 53, line 47, where the symbol for sulphur is
incorrectly given instead of that for liver of sulphur.

The translator has not transcribed Bergman's table of double
attractions (Table I) directly, but on pages 369 to 382 has explained what each of them means. He has misconstrued a few of
them: for instance Scheme 5 means that pure volatile alkali
(ammonia) produces no change in a solution of muriated lime
(calcium chloride). Scheme 22 means that vitriolated vegetable
alkali (potassium sulphate) and nitrated lead (lead nitrate) in
solution react to produce nitre which remains in solution and
vitriolated lead (lead sulphate) which is precipitated. The
translator may have mistaken the symbol for nitre for the rather
similar symbol for common salt, which is usually a circle with a
horizontal bar. However, in Scheme 32 a barred circle with a 'c'
beside it (for *communis*, common) represents common salt.
Similarly Bergman uses the simple symbol for sal ammoniac in
Scheme 44. It would have been more consistent for him to have
used a compound symbol combining the symbols for the acid and
the base involved, as he does in several other places. Bergman is

also inconsistent in using the symbol for a metal alone sometimes, and the symbol for the calx of the metal at other times. In Scheme 52 he uses to mean an alloy of gold and silver the symbol made by combining the symbols for gold and silver by which elsewhere he means platinum.

As we have already seen, Bergman's use of these diagrams to show double decompositions is derived from Cullen and Black; but he carefully avoids anything which might imply a theory about causes or mechanisms. The arrows in Schemes 9–11 merely mean that one substance is partitioned between two others, an original observation of Bergman's made possible by his realization of the importance of the relative quantities of reacting substances, and do not mean that there is a special force acting. The importance of the table is that it is the first systematic collection of a large number of such diagrams, covering a large part of the inorganic chemistry then known and showing more detail than before. A particular innovation is the use of the symbol for water or for heat in the middle of the diagram to show whether the reaction is in the wet way or the dry way.

Bergman's *Dissertation* was widely read and influential. Goethe in 1809 wrote a novel called *Die Wahlverwandtschaften* (Elective Affinities) about a double decomposition of marriages, whose title was clearly inspired by Bergman; but by then it was out of date. Bergman's work had shown not that an enlarged table of attractions could systematize the whole of chemistry but that a reliable table was impossible; his table is almost a *reductio ad absurdum*. Nevertheless, the fascination of the *Dissertation* is that it shows us the highest level of chemical thought immediately before the coming of Lavoisier's New Chemistry, and the end of the phlogiston theory. Bergman's knowledge of chemical phenomena and his skill in interpreting reactions are so great that reading his work one can understand how so many good chemists were content to work with the phlogiston theory for so long. Time and again one feels that Bergman's quantitative exploration of the theory and his thoroughness in working out the implications of it must surely reveal its essential weakness; but his faith in it remains unshaken and it was left to others to question its central assumption. Yet much in the *Dissertation* is original—his use of algebraic symbols, his use of diagrams, his insight into the nature of reactions in solution and his recognition that the result of a reaction might be controlled by the relative quantities of the reagents. Even though a barrier separates the chemistry of Lavoisier and Dalton from the old chemistry of which Bergman was a master, there is a great deal in nineteenth-century theoretical and physical chemistry which can be traced back to him.

A. M. DUNCAN.

NOTES

The standard source of detailed information about chemists and their discoveries is J. R. Partington, *A History of Chemistry*, London, vol. ii, 1961; vol. iii, 1962; vol. iv, 1964. For Bergman's work, see Birgitta Moström, *Torbern Bergman, a Bibliography of his Works*, Stockholm, 1957. Volume I of the collected edition of *Torbern Bergman's Foreign Correspondence*, edited by Göte Carlid and Johan Nordstrom (Stockholm, 1965) is now available.

1. French extracts from the original version in Rozier's *Observations sur la Physique*, 1778, **13** (Supplement), pp. 298–333. French translation, 1788; German translation, 1785; Italian translation, 1801.

2. Newton, *Opticks*, London, 1704 (Dover reprint, New York, 1952), pp. 375–406.

3. Reprinted as Black, *Experiments upon Magnesia Alba*, Alembic Club Reprint N 1, Edinburgh, 1944.

4. See D. McKie and N. H. de V. Heathcote, *The Discovery of Specific and Latent Heats*, London, 1935.

5. H. Cavendish, 'Three Papers, containing Experiments on factitious Airs', *Phil. Trans.*, 1766, lvi, p. 141; H. Cavendish, *Scientific Papers*, London, 1921, Vol. ii, p. 77; Alembic Club Reprint No. 3, 1952.

6. *Collected Papers of Carl Wilhelm Scheele* (trans. L. Dobbin), London, 1931. J. Priestley, *Experiments and Observations on Air*, London, 3 vols, 1774–77. Alembic Club Reprints Nos. 7 and 8, 1947, 1952.

7. E. F. Geoffroy, *Mém. Acad. R. Sci.*, 1718, pp. 202–212. *Hist. Acad R. Sci.*, 1718, pp. 35–7. For a discussion of other tables see A. M. Duncan, 'Some Theoretical Aspects of Eighteenth-century Tables of Affinity', *Annals of Science*, **18**, pp. 177–194 and 217–232.

8. Geoffroy, *Mém. Acad. R. Sci.*, 1720, pp. 20–23.

9. C. E. Gellert, *Anfangsgründe zur Metallurgischen Chymie*, Leipzig, 1751.

10. A. Baumé, *Manuel de Chymie*, Paris, 1763, p. 7.

11. Lavoisier, *Mém. Acad. R. Sci.*, 1782, p. 530.

12. D. A. Rüdiger, *Systematische Anleitung zur reinen und überhaupt applicirten oder allgemeinen Chymie*, Leipzig, 1756, pp. 248 ff.

13. J. P. de Limbourg, *Dissertation sur les Affinités Chymiques*, Liège, 1761, p. 48.

14. J. F. Demachy, *Recueil de Dissertations Physico-chymiques*, Amsterdam, 1774; de Fourcy, Rozier's *Observations sur la Physique*, 1773, **2**, pp. 197–204.

15. P. A. Marherr, *Dissertatio Chemica de Affinitate Corporum*, Vindobonae, 1762.

16. J. Béguin, *Tyrocinium Chymicum*, Paris, 1615, pp. 167–8. See T. S. Patterson, *Annals of Science*, **2**, 1937, pp. 243–298.

17. A. Plummer, *Essays and Observations Physical and Literary*, vol. ii, Edinburgh, 2nd edition, 1771, pp. 381–3.

18. J. Black, *op. cit.*, p. 44.

19. On the manuscripts of Cullen's lectures see W. P. D. Wightman, *Annals of Science*, **11**, 1955, pp. 154–165 and **12**, 1956, pp. 192–205. For a discussion (to which I am greatly indebted) of these diagrams see M. P. Crosland, *Annals of Science*, **15**, 1959, pp. 75–90.

20. J. Black, *Lectures on the Elements of Chemistry*, Edinburgh, 1803, vol. i, pp. 279–80.

21. Anon. [J. P. Macquer], *Dictionnaire de Chymie*, Paris, 1766, vol. i, pp. 49–50.

22. Guyton de Morveau, *Élémens de Chymie*, Dijon, 1777, vol. i, pp. 54, 62 ff. See W. A. Smeaton, *Ambix*, 1963, **11**, pp. 55–64.

23. F. C. Achard, *Mém. Acad. Berlin*, 1776, p. 149.

24. C. F. Wenzel, *Lehre von der Verwandtschaft der Körper*, Dresden, 1777.

25. R. Kirwan, *Phil. Trans.*, 1781, **71**, pp. 7–41; 1782, **72**, pp. 179–236; 1783, **73**, pp. 15–84.

26. J. Elliott, *Elements of the Branches of Natural Philosophy Connected with Medicine*, London, 1782, plate IV.

27. A. F. de Fourcroy, *Mémoires et Observations de Chimie*, Paris, 1784, pp. 308, 438. See also W. A. Smeaton, *Fourcroy, Chemist and Revolutionary*, Cambridge, 1962, p. 31.

28. Guyton de Morveau, *Encyclopédie Méthodique*, Paris and Liège, 1789, Tome I, pp. 466–490.

29. J. L. L. de Buffon, *Histoire Naturelle*, Tome xiii, 'Second Vue de la Nature', Paris, 1765, pp. xii–xvi. Cp. Macquer, *op. cit.*, vol. ii, p. 195.

30. Macquer, *op. cit.*, vol. i, pp. 49–54.

31. A. Baumé, *Manuel de Chymie*, 2nd edition, Paris, 1765 (1763), pp. 8–14.

32. Guyton de Morveau, *Élémens de Chymie*, 1777, vol. i, pp. 79–97.

33. M. P. Crosland, *Historical Studies in the Language of Chemistry*, London, 1962, pp. 142–152, 227–244.

34. Bergman, 'De acido aerëo commentatio,' *Nova Acta Reg. Soc. Sci. Upsal.*, **2**, 1775, p. 108.

35. De Morveau, Lavoisier, Bertholet [*sic*], and de Fourcroy, *Methode de Nomenclature Chimique*, Paris, 1787.

36. Bergman, *Dissertatio Chemica de Diversa Phlogisti Quantitate in Metallis, quam. . . . publice ventilandam sistit Andreas Nicolaus Tunborg, Dalekarlus*, Upsala, 1780.

In some European universities in the eighteenth-century the graduand's dissertation was written by himself; but since a number of dissertations which are said on their title pages to have been defended by a student of Bergman's are included without question by his contemporaries among his works, presumably at Uppsala the professor wrote them.

A

DISSERTATION

O N

ELECTIVE ATTRACTIONS.

By TORBERN BERGMANN,

LATE PROFESSOR OF CHEMISTRY AT UPSAL, AND
KNIGHT OF THE ROYAL ORDER OF VASA.

Tranflated from the Latin by the TRANSLATOR
Of SPALLANZANI'S DISSERTATIONS.

L O N D O N:

PRINTED FOR J. MURRAY, No. 32. FLEET-STREET;
AND CHARLES ELLIOT, EDINBURGH.

M,DCC,LXXXV.

P R E F A C E.

THE ufefulnefs of publications, which, like the prefent Differta- tion, exhibit, from time to time, com- prehenfive views of fcientifical know- ledge, has been fufficiently pointed out by Lord BACON, whofe dictates upon this fubject, as upon others, have been amply confirmed by experience. The Tranflator, therefore, at firft thought, that every purpofe of the Englifh reader would be abundantly ferved by a faith- ful tranflation of this admirable manual of theoretical Chemiftry. His duty plainly forbad him to alter or fupprefs any thing; and his reverence for the great author deterred him from the thought of making any addition. But fome time has elapfed fince the tafk of

mere

mere tranflation was completed ; in the
mean time, chemical inveftigation being continued with univerfal ardour,
new facts were brought to light, and
new theories propofed, fome of them in
books not likely to fall into the hands
of every reader.   Hence it feemed almoft a matter of neceffity to add fome
annotations.   The Tranflator   now
wifhes, for the convenience of the
reader, that they had been fubjoined to
the pages to which they refer, though,
for his own fake, he is not forry that
they are thrown back as far as poffible.
This accidental circumftance of their
fituation has led him to be more dif-
fufe than he would otherwife have
been.   The notes could not, by their
intrufion on the reader's eye, divert his
attention from the author ; and why
fhould any thing which was ufeful, and
perhaps

perhaps inacceſſible to many, be with-
held, when it had any connection with
the ſubject ?

Two ſets of Tables are ſubjoined. It
was thought that many readers would
be diſſatisfied with the chemical Characters alone, eſpecially as the former
edition of the Tables has been already
publiſhed in words. To ſupprefs the
ſigns entirely, ſeemed improper ; for
they are ſo convenient, that every ſtu-
dent of chemiſtry ought to make him-
ſelf familiar with them.   Beſides, as
moſt Chemiſts will wiſh for a ſet to
ſtand always open for inſpection, the
two ſets will ſcarce be thought ſuperfluous by any.

Every man who delights in paying
the reſpect that is due to genius, learning,

ing, and induftry, will hear with pleafure, that the life of our confummate
philofopher has been promifed us by his
excellent friend, Mr Scheele. The following teftimony of regret, on account
of his premature death, appeared foon
after that event ; an event which thofe
who, by comparing what he did for
chemiftry with the fhort time during
which he applied to it, fhall become
fenfible of what he would have accomplifhed in a long life, can alone ade-
quately lament.

Pertriftem

Pertriftem adferre nuncium oportet,
Poft exactos mortali vita annos xlix,
Ereptum effe Patriæ,
Cultiori orbi, bonorumque omnium amplexibus,
Virum longè celebratissimum,
T O R B E R N U M   B E R G M A N N.
Chem. metallurg. et pharmaceut. in
Academ. Upsaliensi Professorem ;
Equitem auratum Reg. Ord. de Wasa ;
Acad. Imp. N. C. ; Regiarum Academiarum
Et Societatum Paris. Medicæ Paris.
Montispess. Divionens. Upsal.
Stockh. utriusque Londin. Goetting.
Berolin. Taurin. Gothoburg. Lund.
S o d a l e m.

Id quod accidit d. viii. Julii anni 1784,
Dum ad acidulas Medvicenfes, in Oftrogothia,
Adflictæ dudum valetudini quærebatur folatium,
Lugent per univerfam Suecogothiam
Optimarum fcientiarum Patroni et Cultores ;
Necdum inveniunt
Quem tanti viri defiderio modum ponant.
Superftitibus, autem, cognatis et amicis,
Hoc denique lacrimabile officium relictum eft,
Ut, mœftiffimæ nomine viduæ,
Fautores in exteris gentibus et Confortes ftudiorum,
Quæ naturalis omnis fcientiæ vir peritiffimus
Excoluit,
De communi clade certiores faciant,
Proque fuâ adeo agant parte,
Ut juftiffimi luctus æquè latè fentiatur pietas,
Ac Bergmanniorum exiftimatio meritorum
Jam diu inclaruit,

Upsaliæ d. xvi. Julii ⎱
1 7 8 4.   ⎰

CONTENTS.

DIS-

DISSERTATION, &c.

Jamne vides igitur magni primordia rerum
Referre, in quali fint ordine quæque locata,
Et commifta quibus dent motus accipiantque.

LUCRETIUS.

THIS Differtation was firft printed in 1775, in the third volume of the New Upfal Tranfactions. It was afterwards tranflated both into German and French. The two annexed Tables, which exhibit the fingle and double attractions, were again engraved in London, by the care of Dr Saunders, for the ufe of thofe who attended the lectures which he read in conjunction with Dr Keir. The fame year, Mr More, fecretary to the Society for the Encouragement of Arts and Manufactures, publifhed the table of elective attractions, on a large fheet, fubftituting Englifh words in the place of the figns.

ON

ELECTIVE ATTRACTIONS,

I.

There feems to be a Difference between Remote and Contiguous Attraction.

IT is found by experience, that all fubftances in nature, when left to themfelves, and placed at proper diftances, have a mutual tendency to come into contact with one another. This tendency has been long diftinguifhed by the name of *attraction*. I do not purpofe in this place to inquire into the caufe of thefe phænomena ; but, in order that we may confider it as a determinate power, it will be ufeful to know the laws to which it is fubject in its operations,

though

though the mode of agency be as yet unknown.

It has been fhewn by Newton, that the great bodies of the univerfe exert this power directly as their maffes, and inverfely as the fquares of their diftances. But the tendency to union which is obferved in all neighbouring bodies on the furface of the earth, and which may be called *contiguous attraction*, fince it only affects fmall particles, and fcarce reaches beyond contact, whereas remote attraction extends to the great maffes of matter in the immenfity of fpace, feems to be regulated by very different laws; it feems, I fay, for the whole difference may perhaps depend on circumftances. Confidering the vaft di-ftance, we may neglect the diameters, and look upon the heavenly bodies, in moft cafes, as gravitating points. But conti-guous bodies are to be regarded in a very different light; for the figure and
fituation,

fituation, not of the whole only, but of the parts, produce a great variation in the effects of attraction. Hence quantities, which in diftant attractions might be neglected, modify the law of contiguous attraction in a confiderable degree; and, moreover, the great power of our globe on all occafions influences and difturbs it. This force may therefore produce wonderful variations in the effects, according to circumftances. But as we are by no means able to afcertain the figure and pofition of the particles, it remains that we determine the mutual relations of bodies with refpect to attraction in each particular cafe, by experiments properly conducted, and in fufficient number.

As contiguous attraction fcarce extends beyond contact, it is obvious, that the former will be more intenfe in the fame body, the more the latter is increafed. Hence, in the following obfervations,

fervations, many inftances occur, which prove that a body has greater power in a liquid than in a folid ftate, and ftill greater when it is refolved into vapour. See, in particular, what is faid of marine acid, with refpect to phlogifton, in the forty-feventh paragraph; and of the matter of heat, in the forty-eighth.

In this differtation, I fhall endeavour to determine the order of attractions according to their refpective force ; but a more accurate meafure of each, which might be expreffed in numbers, and which would throw great light on the whole of this doctrine, is as yet a defi‧deratum *.

SEVERAL

---

* Mr Morveau, I believe, firft determined and expref‧fed in numbers the cohefion of quickfilver to fome of the metals. Mr Achard afterwards publifhed a large table, in which the cohefive force of many bodies is inveftigated both by calculation and experiment. Mr Kirwan very lately

SEVERAL fpecies of contiguous attraction may be diftinguifhed. I fhall here briefly mention the principal. When homogeneous bodies tend to union, an increafe of mafs only takes place, the nature of the body remaining ftill the fame; and this effect is denominated *the attraction of aggregation.* But heterogeneous fubftances, when mixed together, and left to themfelves to form combinations, are influenced by difference of quality rather than of quantity. This we call *attraction of compofition*; and when it is exerted in forming a mere union of two or more fubftances, it receives the name of *attraction of folution* or *fufion,* according as it is effected either in the moift or the dry way. When it takes place between three refpectively, to the exclufion of one, it is faid to be a *fingle elective*

---

lately began to meafure the attractions , by the diminution of bulk that is produced by union ; for he is of opinion, that the caufe and quantity of contraction is to be fought in the force of attraction.

elective attraction; when between two compounds, each confisting of only two proximate principles, which are exchanged in confequence of mixture, it is intitled *double attraction*. I am particularly to confider the two laft fpecies.

II.

Single Elective Attractions.

SUPPOSE *A* to be a fubftance for which other heterogeneous fubftances *a, b, c,* &c. have an attraction; fuppofe, further, *A*, combined with *c* to faturation, (this union I fhall call *Ac*), fhould, upon the addition of *b*, tend to unite with it to the exclufion of *c*, *A* is then faid to attract *b* more ftrongly than *c*, or to have a ftronger elective attraction for it; laftly, let the union of *Ab*, on the addition of *a*, be broken, let *b* be rejected, and *a* chofen in its place, it will follow, that *a* exceeds *b* in attractive power,

and

and we fhall have a feries, *a, b, c,* in refpect of efficacy. What I here call attraction, others denominate affinity ; I
fhall employ both terms promifcuoufly
in the fequel, though the latter, being
more metaphorical, would feem lefs proper in philofophy.

GEOFFROY, in 1718, firft exhibited at
one view the feries of elective attractions, by arranging in a table the chemical figns, according to a certain order ; but this admirable contrivance,
while it is commended by fome, is bla-
med by others ; one party contending,
that affinities are governed by fixed
laws, and the other affirming, that they
are vague, and to be afcribed to circumftances alone.

Now, fince all chemical operations
confift either in analyfis or fynthefis,
compofition or decompofition, and both
the one and the other depend on attraction,

tion, it will certainly be of great importance to determine this difpute. Let us not then lightly, and on account of one or two irregularities, perhaps ill underftood, reject the whole doctrine, but let us rather proceed in our examination with caution and care. Should we even at laft find that attractions depend on circumftances, fhall we therefore conclude, that it will be ufelefs to know the feveral conditions that forward or impede or difturb them? By no means, but rather that it will be of extenfive utility. There does not exift in all nature a fingle phænomenon but what is fo connected with certain conditions, that when they are abfent, the phænomenon fhall either not appear, or be varied occafionally. It is of confequence to fcience, that the changes and the combination of caufes in every operation fhould be accurately known, as far as a knowledge of them

is

is attainable ; and the utility of a ſtrict inquiry into attractions will, I hope, clearly appear from many inſtances in the following pages.

But if, on the contrary, a fixed order does really take place, will it not, when once aſcertained by experience, ſerve as a key to unlock the innermoſt ſanctuaries of nature, and to ſolve the moſt difficult problems, whether analytical or ſynthetical ? I maintain, therefore, not only that the doctrine deſerves to be cultivated, but that the whole of chemiſtry reſts upon it, as upon a ſolid foundation, at leaſt if we wiſh to have the ſcience in a rational form, and that each circumſtance of its operations ſhould be clearly and juſtly explained. Let him who doubts of this conſider the following obſervations without prejudice, and bring them to the teſt of experiment.

III.

III.

Whether the Order of Attractions be constant.

This question can only be properly
answered from what follows.   But let
us now slightly consider whether a constant series, such as is mentioned in the
last paragraph, is to be expected.   Does
a expel *b*, and *b a* reciprocally, according to circumstances ?   Does *c* perchance
expel *a*, while it always gives way to *b* ?
Let us consult Experiment, the oracle
of nature, with due care and patience,
and we shall doubtless find the proper
clue to guide us out of this labyrinth.

I am far from approving of those general rules which affirm, that earths
and metals are in all cases precipitated
by alkalis, and metals by earths, for
they are often fallacious.   We have,
however, many particular observations,
which,

which, when every thing is properly difpofed, never miflead. We know, for inftance, that volatile alkali is diflodged by fixed alkali and pure calcareous earth; that quickfilver and filver are precipitated from nitrous and vitriolic acids on the addition of copper, which is again feparated by iron. Silver, quickfilver, and lead, which were called the white metals by the ancients, are feparated from the nitrous acid both by the vitriolic and marine. Do not thefe, and other facts long fince known, fhew, that there prevails a conftant order among thefe feveral fubftances? Many other clear proofs occur in the explanation of the new table of attractions, which I fhall referve for their proper places, (XII.—LXX.). The difficulties, when clofely examined, difappear; and none has yet, as far as I know, been pointed out which is really inconfiftent with a continued feries. But fhould there occur in this, as in other branches

of

of natural philofophy, a few phænomena, which appear to deviate from the
ordinary track, they fhould be confidered as comets, of which the orbits cannot yet be determined, becaufe they
have not been fufficiently obferved.
Repeated obfervations, and proper experiments, will in time difpel the dark-
nefs.

That the effect of three fubftances
mixed together may appear at one view,
I have contrived a way of reprefenting
it by fymbols.   It will be proper to illuftrate it by an example.

Scheme 20. Pl. i. exhibits the decompofition of calcareous hepar by the
vitriolic acid.   On the left fide appears
the hepar, indicated by the figns of its
proximate principles united ; but within the vertical bracket thefe principles
are feen feparate, one above the other.
On the right, oppofite the calcareous
earth,

earth, is placed the fign of vitriolic
acid ; in the middle ftands the fign of
water, intimating that the three furrounding bodies freely exercife their attractive powers in it.   Now, as vitriolic acid attracts calcareous earth more
forcibly than fulphur does, it deftroys
the compofition of the hepar ; the extruded fulphur being by itfelf infoluble
falls to the bottom, which is fignified
by the point of the lower horizontal
half-bracket being turned downwards ;
and as the new compound, vitriolated
calcareous earth (gypfum), alfo fubfides,
unlefs the quantity of water be very
large, the point of the upper bracket is
likewife turned downwards.   The complete horizontal bracket indicates a
new combination, but the half-bracket
ferves merely to fhew by its point whether the fubftance from which it is
drawn remains in the liquor, or falls to
the bottom.   The abfence of horizontal brackets indicates that the original
compound

compound remains entire. Such a
combination only as continues unaltered can have a place on the right fide,
for if it be likewife decompofed, a new
cafe arifes, which will be noticed hereafter, (V.). Thofe operations which are
performed in the dry way, are diftinguifhed by the character of fire, which
is placed in the middle.

IV.

A Difference in the Degree of Heat fometimes
produces a Difference in Elective Attrac-
tions.

THE only external condition, which
either weakens or totally inverts the affinities of bodies fubjected to experiments, is the different intenfity of heat.
But this caufe can only operate in cafes
where the fame temperature renders
fome

fome bodies remarkably volatile in comparifon of others.

SUPPOSE *A* to be attracted by two
other fubftances ; and let the more
powerful act at the ordinary temperature with the force *a*, the weaker with
the force *b*: fuppofe, at the fame time, the
former to be the more volatile ; let its
effort to arife be expreffed by *V*, and that
of the other by *v*. When thefe three fubftances are mixed together, the ftronger will attract *A* with a force $=a-b$;
but fhould the heat be gradually raifed,
this fuperior force will be more and
more diminifhed ; and as *V* will increafe fafter than *v*, we fhall at laft
have $a-b=V-v$. This ftate of equilibrium will be immediately deftroyed
by the fmalleft addition of heat ; and
thus *b*, which was before the weaker,
and incapable of producing any effect,
will now prevail. If the other fubftance
be entirely of a fixed nature, $v=$o, and
the

the cafe will be fimpler. Many in
ftances of this nature will hereafter
occur.

HENCE, I think it in general obvious, that thofe are the genuine attractions, which take place when bodies
are left to themfelves : too high a degree of heat is an external caufe, which
forcibly weakens the real affinities more
or lefs, nay, in fome cafes, even totally
alters them. Since, however, many operations cannot be carried on without
the aid of heat, and the power therefore of this moft fubtile fluid is highly
worthy of being obferved, I think the
table of elective attractions ought to be
divided into two areas ; of which the
upper may exhibit the free attractions,
that take place in the moift way, as
the expreffion is ; and the lower, thofe
which are effected by the force of heat.
This may eafily be done, fince we are
as yet unacquainted with any other external

ternal condition which deferves here to
be taken into the account; if the internal conditions ever caufe any deviation,
it is either only apparent, or elfe a real
change in the nature of the fubftances
is produced. It cannot indeed be denied, that volatile bodies are actually
changed by a combination with the
matter of heat; but the change is of
fhort duration, as it totally difappears
on refrigeration, though not till after
the defired decompofition has been effected.

It is hence evident, what opinion we
are to form concerning the various arguments brought againft the conftancy
of affinities, from the diftillation, fublimation, or fufion of mixtures: fuch
fometimes is the efficacy of heat, that
ftrong digeftion, or even that degree of
warmth which is produced by the combination of certain fubftances, is fufficient to difturb the ufual order.

V.

V.

*Apparent Irregularities from a double At-
traction.*

NOTWITHSTANDING the trite pro-
verb, That no rule is without exception, I do not therefore think that rules
are to be rejected; but the exceptions
fhould be properly inveftigated, and
the rules thus be reduced to their juft
amount. Thofe, however, which are
now to be confidered, do not come under the denomination of exceptions;
for four fubftances exert their action, a
very different cafe, and more complicated, than where three only are concerned. Many inftances ufually adduced in refutation of a regular feries of
affinities belong to this head; and
though it is faid that reciprocal decompofition is evidently fhewn by them, a

clofer

clofer examination will diffipate the il-
lufion.

GEOFFROY's table intimates, that fixed alkalis adhere to acids more ftrongly
than calcareous earth, and with great
propriety, though fome would reprefent
this truth as an abfurdity. They drop
a folution of chalk in nitrous acid into
a folution of vitriolated tartar, upon
which a precipitation of gypfum immediately takes place ; a clear proof, as
they think, of the fuperior attraction of
the calcareous earth. But it fhould be
obferved, that not even quicklime,
when added to a folution of vitriolated
tartar, (Scheme 2.), produces any decompofition ; but, on the contrary, if vitriolic acid be dropped into the folution of
calcareous earth, a precipitation of gypfum will follow, (Scheme 16.): Hence
it is obvious to which the ftronger elective attraction is to be attributed. When
the chalk has been previoufly diffolved
in

in some mineral acid, four substances
come into action; and now the earth,
aided by the acid which is combined
with it, effects what before it was not
able to accomplish. To render this
more evident, let us confider the 21st
Scheme, in which, on the left hand, the
neutral salt, known by the name of vitriolated tartar, is indicated by the sign
of vegetable fixed alkali placed near that
of vitriol. I read it in this manner : Vitriolated vegetable alkali, (that is, saturated with the acid obtained from vitriol), a denomination by which the
proximate principles are known ; and
these also appear under their proper
signs within the adjacent vertical bracket. On the right, muriated lime (saturated with the marine acid) is likewise
reprefented by fymbols, as alfo its proximate principles within their proper
bracket. Whilft therefore vitriolated
tartar and lime faturated with the acid
of falt are mixed together in water,
(which

(which is expreffed by the fign of water in the middle of the Scheme), the fame thing happens as if we were to mix certain portions of vegetable fixed alkali, vitriolic acid, marine acid, and pure lime in water : thefe four fubftances furround water in the figure, and are fo to be placed, that the two acids fhould never be in the fame horizontal line.

WE have the fubftances that were combined before mixture difpofed in a vertical pofition ; and, in order to break the combination, there is neceffarily required a greater fum of attraction between thofe which are horizontally, than thofe which are vertically oppofite to each other : and fuch is the prefent cafe; for although the vitriolic acid attracts the fixed alkali more ftrongly than it does the lime, yet, upon the addition of muriatic acid, which at once folicits the alkali, and diminifhes its cohefion with the vitriolic acid, the attraction between

the

the fixed alkali and the marine acid,
together with the attraction between
the vitriolic acid and the lime, make together a greater fum than the attraction
between the fixed alkali and the vitriolic acid, together with that between the
muriatic acid and the lime.

The horizontal brackets include the
new combinations : by the apex of the
lower turned downwards, fubfidence is
denoted ; but the apex of the upper
pointing upwards, fhews that the combination contiguous to it remains in the
liquor, until a certain portion be carried
off by evaporation. Thus then is the decompofition in queftion effected. That
I may more quickly difpatch the other
inftances, I fhall once for all obferve,
that I always exprefs double falts (fuch
as confift of two proximate principles)
by the fign of the bafe placed on the
left, near the fign of that falt from
which the acid is ordinarily expelled,

which

which acid is at prefent combined with
the bafe : in the fame way I defign the
others, confifting of two principles intimately united.

MAGNESIA, and feveral of the metals, refemble lime in this refpect. See
Scheme 22. Table 1. contains fixty-four
fchemes, which exhibit the events of
122 experiments ; for No. 24. not only
fhews, that nitrated mercury is decompofed by feret fal ammoniac, but alfo
that flaming nitre and vitriol of mercury will bear to be mixed without fuffering any change ; and, in general, if the
fubftances reprefented on the right and
left fide make a mutual exchange of
their principles, we may judge that
thofe which ftand above and below undergo no feparation on mixture. In
Schemes 21. and 23. it appears that
vitriolated tartar and muriated lime
change their principles, but not digeftive falt and gypfum. Nos. 1.—20. exhibit

hibit the fingle, 21.—40. the double free
attractions. The following are brought
about by the acceffion of fire : 41.—47.
and 55.—58. by diftillation ; 48.—50.
and 59.—62. by fublimation ; and, laftly, 51.—54. 63. and 64. by fufion in a
crucible. An explanation of the new
characters is found in XI.

CHEMISTS, in determining the fingle
elective attractions, are often deceived
by double attractions. The phofphoric
acid, as I fhall hereafter fhew, (XXXIII.),
attracts lime more powerfully than fixed alkali ; yet the contrary is afferted,
fince aerated alkali, by means of a
double affinity, precipitates phofphorated lime. Even cauftic fixed alkali,
which feems a ftronger proof, caufes a
precipitation : neverthelefs, if the fuperior attraction is deduced from this
alone, the conclufion will be erroneous;
for the alkali only takes away the excefs of acid which is requifite for folution,

tion, and hence the phofphorated lime
falls to the bottom, (IX.). The difference between the action of alkalis and
abforbent earths, when faturated with
aerial acid, and when deftitute of it, has
been explained in my effay on that acid,
and may therefore be omitted here *.
What is there faid of the volatile alkali,
is expreffed in Scheme 36. and clearly
fhews the reafon why cauftic volatile alkali feems to act more weakly
than the mild, which otherwife would
be abfurd. It may now, in general, be
obferved, that alkaline and earthy fubftances are to be confidered as double,
and by no means as fimple falts, except
in their cauftic ftate, and then I call
them pure. See Schemes 1.—8. 32.—37.
51. 62. and 63.

THE precipitation of metals, diffolved
in acids, by other metals, is never the
effect of fingle attraction; for during
the

* Op. vol. i. p. 49.

the folution, a quantity of phlogifton is extricated, as I have clearly, I hope, fhewn elfewhere *. Since therefore metallic folutions are of fuch a nature, that they cannot reftore what they hold diffolved to its metallic fplendour without the acceffion of a new portion of phlogifton, it is felf-evident, as well as conformable to experiment, that this cannot be effected by the addition of calces. If therefore ochre be put into a folution of vitriol of copper, no copper will be precipitated; but iron added to the folution is foon obferved to be covered with a cupreous pellicle; for it yields part of its phlogifton, which is neceffary to the reduction of the copper, and by this means becomes itfelf foluble without the emiffion of any inflammable air, (Scheme 39.). Silver diffolved in the nitrous acid is in like manner calcined; as appears from the red vapour, phlogiftic fmell, and various

rious

* Op. vol. ii. p. 354. vol. iii. p. 134.

rious other evident figns, and therefore cannot be precipitated by the calx, though it may by regulus of copper, (Scheme 38.). The fame obfervation is applicable to gold and the other metals; for in whatever way they be feparated, provided they can acquire no phlogifton, they appear calcined, and really are fo: the only difference confifts in this, that they are unequally dephlogifticated, and that the noble metals can recover their lofs by fufion alone in ignited veffels, whereas the ignoble ones require an addition of phlogifton. But more of this hereafter.

In many other cafes, where a fingle elective attraction is commonly thought to take place, it is really double, on account of the prefence of phlogifton. Let us confider an inftance of this in the diftillation of butter of antimony from a mixture of corrofive fublimate and regulus of antimony. We may obferve, in the

the firſt place, that neither mercury nor
regulus of antimony are ſoluble in the
marine acid, unleſs they are firſt deprived of a certain portion of their inflammable principle. From this conſidera-
tion, the proceſs is eaſily explicable
from a double attraction: the calx of
mercury in corroſive ſublimate is revivified by that phlogiſton which the regulus muſt loſe in order to become ſoluble in marine acid, (Scheme 58.). The
baſis of corroſive ſublimate, indeed, as
well as the calces of the noble metals,
recovers its reguline ſtate in heat alone,
by attracting what is wanting to this
ſtate, through the ignited veſſels; but
this operation requires a far ſtronger
fire than the diſtillation of butter of antimony, in which the reduced mercury
riſes without any very ſtrong heat.
With white arſenic and corroſive ſublimate no decompoſition is effected, ſince
the phlogiſton neceſſary for the reduction of the mercury is wanting; but
this

this operation fucceeds with orpiment,
which abounds with phlogifton.

VI.

*Apparent Exceptions from a fucceffive Change
of Subftances.*

IF either of the fubftances employed
fhould change its properties, its attractions will, doubtlefs, be liable to alteration. This may be illuftrated by many
examples. It had been long known,
that nitrous acid is capable of dif-
lodging the marine from an alkaline
bafis ; but Margraaf was the firft who
obferved that the latter expels the former. This phænomenon, unlefs we are
acquainted with the nature of marine
acid, eludes all explanation; but now
this is known, the problem is eafily fol
ved. The nitrous acid expels the marine, by means of a fingle elective attraction,

traction, (Scheme 42.), but phlogiston
enters into the composition of the marine acid, and is attracted from it by
other acids, especially by the nitrous,
(XVII.), which, even though faturated
with vegetable alkali, strongly attracts
the inflammable principle; for nitre
kept in a state of ignition for an hour
or two, remains perfectly neutral,
whence it is plain, that all the acid remains, but it is become so much weakened, by being phlogisticated, that it
may be expelled by concentrated vinegar, (XXXVII.). Hence it follows, that
part of the marine acid that is poured
upon the nitre, yields its phlogiston, in
consequence of the heat applied, to the
nitrous acid, which in this state is expelled by that part of the marine which
is not yet decomposed, (Scheme 55.).
That this is the true explanation, appears from the nature of the thing, the
neceffary proportions, and the matter
collected in the receiver, which is found

to

to confift of phlogifticated nitrous acid
and marine acid, both dephlogifticated
and in its ordinary ftate.

On the fame principle, white arfenic
is capable, by diftillation, of decompofing thofe neutral falts which contain
the nitrous, but not thofe which contain the marine acid.   White arfenic is
nothing but a fort of fulphur, confifting of the arfenical acid, and a certain
portion of phlogifton, (XX.).   In this
cafe, therefore, four fubftances come
into action, (Scheme 56.) ; and as the
nitrous acid ftrongly attracts phlogifton,
and its connection with its bafis is much
weakened by the acceffion of that principle, the acid of arfenic is capable of
expelling it : but the marine acid, already containing the inflammable principle, refufes to combine with a larger
portion of it, it therefore remains
unchanged, and the arfenical acid
has

has no power againſt the ſtronger, (Scheme 57.).

I have long ſince obſerved, that the noble metals, though they reſiſt the force of fire ſo obſtinately, may be more or leſs calcined by ſolution in acids ; and it will now be proper to add ſome-thing concerning iron, which, above all, ſeems liable to this change, eſpecially concerning its ſolution in the vitriolic acid. In the firſt place, we remark that a portion of phlogiſton flies off du-ring ſolution in the inflammable air ; next, if the ſaturated ſolution, which is of a green colour, be filtered, and kept in a full and cloſe phial, it will remain clear ; whereas, if it be expoſed to the air in an open veſſel, it will gradually, but conſtantly, depoſit ochre, a phæno-menon which ariſes from two cauſes. For vital and reſpirable air attracts phlogiſton ſo forcibly as infenſibly to diminiſh that which is contained in the

ſolution

folution of iron. Such alfo is the nature of the vitriolic ·acid, that it will diffolve fo much the lefs iron, the more deftitute the metal is of phlogifton; whence it follows, that a quantity of acid fufficient for the folution of iron but little dephlogifticated, gradually becomes infufficient in proportion as the phlogifton is feparated, and therefore earth of iron muft fall to the bottom, which, however, again difappears on the addition of frefh acid. This decompofition is much accelerated by heat, and efpecially by boiling, and at laft the green colour is changed to a dark red, and the whole folution affumes the nature of an ultimate ley, which is incapable of cryftallization, as has been admirably fhewn by Mr Monnet. Succeffive boiling, however, and cooling, bring about the dephlogiftication fooner than continued boiling alone.

THIS

This ultimate lixivium, however, may be reduced by the addition of a little vegetable alkali, to white aftringent cryftals, like thofe of alum, whence fome have been led to believe, that the tranfmutation of vitriol into alum is fully proved ; but this falt, when diffolved, may be totally changed into Pruffian blue, by means of phlogifticated alkali, and, if common alkali be employed, affords nothing but ochre, and not a particle of alum, provided the vitriol be made by diffolving iron ; that indeed, which is extracted from pyrites, often contains alum, fince clay frequently enters into the compofition of pyrites.

With thefe truths in view, it is eafy to anfwer, if any one fhould think of proving the reality of reciprocal decompofition by the cafe of alum, in which the acid feems to part with the earthy bafis upon the addition of iron filings, and take

up

up the metallic, while, on the contrary, clay is diffolved in the laft lixivium of vitriol, and precipitates the ochre.
In the firft cafe, we have iron combined with its proper portion of phlogifton, which, when it is put into a folution of alum, is taken up only by the
excefs of acid, which adheres more
loofely than the faturating portion.
And the abundant acid being taken
away, the clay, being exactly faturated,
falls down infoluble. Thefe phænomena have therefore been hitherto ill underftood, for at the precife point of faturation, clay is not precipitated by iron.

In like manner, copper yields the
acids to iron, while it attracts them
from crocus martis. Hence the decompofition of vitriol of iron by copper,
detected by Mr Margraaf, is eafily explicable; it is by no means reciprocal,
for copper precipitates iron only when
it is dephlogifticated beyond a certain
limit.

In

In an open veſſel, the inflammable principle is eaſily ſeparated, eſpecially by means of heat ; hence a ſolution of vitriol of iron muſt neceſſarily change in its nature almoſt every moment to a certain point, below which it is impoſſible to proceed in this way. It is the more ſurpriſing, that copper ſhould be ſo eaſily diſſolved in this experiment, ſince it is certain, that this metal is with great difficulty diſſolved in vitriolic acid, unleſs it be in ſome meaſure calcined beforehand. But in the preſent inſtance, the earth of the iron attracts the phlogiſton of the copper, which it ſoon loſes again on the application of heat. The ſolubility of the calx, as well as the regulus, of copper, in a boiling ſolution of martial vitriol, is a clear proof of this conjecture. Concentrated vitriolic acid indeed attracts copper, when aſſiſted by a ſufficient heat ; but let it be remarked, that the vapours which then ariſe, are the phlogiſticated

gifticated vitriolic acid, which fhews,
that a portion of the inflammable principle is carried off. Moreover, how
the precipitations of metals from acids,
by means of other metals, are to be
properly underftood, will be explained
more at large hereafter.

VII.

Apparent Exceptions from Solubility.

IT fometimes happens that no decompofition appears at firft, though it really
takes place. Fixed mineral alkali united with the acids to faturation, and
diffolved in water, remains in the limpid folution on addition of pure vegetable alkali, nor is any congrumation
or precipitation to be obferved. Hence
chemifts of great name have concluded,
that the vegetable does not exceed the
mineral alkali in attractive power ; but
let

let us suppose for a mo: ent, that the
mineral alkali is dislodged, should the
solution become turbid? By no means,
for the fossil alkali is soluble of itself,
and cannot therefore impair transparen-
cy. From this phænomenon alone,
therefore, nothing certain is deducible;
but let the solution be evaporated, and
there will be found at last uncombined
mineral alkali, separately cryftallized,
and befides, vitriolated tartar, if Glauber's salt, cubic nitre, if quadrangular
nitre, and digestive salt, if sea salt was
subjected to the experiment, (Scheme 3.).

Some metals precipitated from acids,
by too much alkali, soon disappear again
in consequence of solution. Platina
and gold can scarce be precipitated in
such a manner that the solutions will
not remain tinged. Zinc, copper,
nickle, and cobalt, afford no precipitate
when an excefs of volatile alkali is used,
unlefs

unlefs they are contaminated by fome heterogeneous fubftance.

There is yet another veil which often hides decompofitions, as when the fubftance expelled from its combination is capable of diffolving the new compound, or at leaft does not hinder the water of folution from doing fo. This happens not unfrequently when the acids of nitre and falt are expelled by thofe of greater power. Thus the vitriolic takes magnefia from the marine, but in fuch a way as totally to efcape the notice of the fenfes ; for the extricated acid, fharpening the water of the folution, immediately takes up the vitriolated magnefia, which is indeed otherwife of very eafy folubility, and therefore no figns of it appear, until, by a fpontaneous evaporation, the menftruum is fo much diminifhed as to be incapable of diffolving the whole. Several inftances of this kind occur in the

the following paragraphs, and it is at the fame time fhewn how this impediment may be removed.

To this head we may alfo refer precipitations happening from a fubtraction of the water of folution, upon the addition of a fubftance which, though it does not change the former combination, yet deprives it of its water, when there is not enough to diffolve both; whence the compound fuddenly concretes into fmall cryftals, nearly in the form of a precipitate. This happens when a proper quantity of concentrated vitriolic acid is poured into faturated folutions of vitriolated tartar, alum, corrofive fublimate, and other falts, not eafily foluble in water, though their proximate principles cannot be parted by vitriolic acid. Vitriol of mercury is really decompofed by marine acid, which attracts the calx of the metal, and carries it down along with it to the

bottom

bottom for want of water : but concentrated vitriolic acid alfo, when poured into a folution of corrofive fublimate, throws down in its turn a white powder ; whence fome have immediately inferred that kind of decompofition which is generally called reciprocal; but the latter precipitate is found, on examination, to be nothing elfe than true corrofive fublimate deprived of the water of folution. Fixed vegetable alkali, particularly when dry and cauftic, produces like effects in the fame circumftances ; I mean, when the bafis attracts the acid with equal or greater force than the alkali employed.

These precipitations are feldom complete ; for fomething generally remains diffolved in the liquor.

Finally, let me notice thofe anomalous phænomena which depend on apparent folubility. Liquor of flints, as

is

is well known, contains filiceous earth, diffolved in water by means of fixed alkali. Upon dropping in an acid, the filiceous earth ought to be precipitated, as really happens, unlefs the liquor be diluted in twenty-four times its weight of water, or more ; in this cafe, no cloudinefs is perceptible, though even more acid be affufed than is neceffary for faturating the alkali. We have here an appearance of folution ; but the truth is, that the filiceous particles are fo difperfed in the abundance of water, that they cannot fubfide on account of the great proportion of their furface to their weight. As I have elfewhere explained this at greater length*, I need only give a fhort view of the matter here. I fhall only add, that the earth cannot be diffolved by the neutral falt that is formed ; for the filiceous particles fall down on ebullition, (which increafes folubility upon other occafions).

* Op. vol. ii. p. 36.

fions), in confequence of the diminifhed denfity of the liquor by heat. Should any one object, that, for the fame reafon, other earths ought alfo to be fufpended, he will readily find, on more mature confideration, that, 1. no other of the primitive earths is foluble in alkali, except the argillaceous, which is alfo foluble in acids, fo that there can be no excefs either of the one or the other without folution. 2. Calcareous earth is not precipitated from acid menftruums in a vifible form by cauftic fixed alkali well diluted, and gradually dropped in, till paper dyed with brazil wood is changed to a blue, provided the folution be firft diluted with fifty times its bulk of diftilled water. The liquor remains clear for a time, but by degrees acquires a film on the furface. The folution employed ought to contain more calcareous earth than an equal quantity of the ftrongeft lime-water, otherwife it might be faid, that the cal-

careous

careous earth is really diffolved in the
clear alkalifed folution, and not fufpended.   If a fingle drop of aerated alkali
be dropped into the clear folution without agitation, white clouds immediately
come into view ; but if an equal drop
be added, while the whole mafs is fha-
ken, no cloudinefs enfues, for the motion prevents the coalefcence of the feparated molecules.   Here then we have
calcareous earth certainly fufpended,
and all ground of contradiction, I hope,
removed.

VIII.

*Exceptions from the Combination of three
Subftances.*

THERE are fome fubftances of fuch a
nature, that three are capable of uniting without the exclufion of any one.
The combination of two attracts a determinate

terminate quantity of a third, and fometimes of more, with fo much force, that
they become very clofely connected,
and are fcarce to be feparated by any
art. This inconvenience more efpecially attends the dry way; for the earths
mixed either with one another, or with
falts, melt together without exclufion,
which is alfo true of moft of the metals.
A is not indeed attracted by *a* and *b*
with equal force, but rather *A a* unites
with *b*, or *A b* with *a*, which yet is no
reafon why *a* fhould not exceed *b* in attraction, though the particular nature
of the combinations prevents exclufion.
Volatile alkali, marine acid, and the
calx of quickfilver, volatile alkali, vitriolic acid, and magnefia; iron, vitriolic acid, and magnefia, not to mention
other inftances, adhere fo clofely in determinate proportions, that they cannot
be feparated by cryftallization, and not
eafily in any other way.

THIS

THIS alfo holds with refpect to four ingredients, as borax with tartar, vitriolated magnefia with common falt, gypfum with common falt, and many others. To this head alfo belongs liver of fulphur formed in the dry way by vitriolated tartar and powder of char-coal, as in this cafe the phlogifton is firft conceived to feparate the acid, and generate fulphur, which then is diffolved in the alkali, and yields hepar; it may feem that the newly formed particles of fulphur can fcarce perfift in fo great an heat, without either being fublimed or confumed, but the new compound is formed almoft in the fame moment.

FROM this property of certain fub-ftances, peculiar phænomena often arife. Should any one attempt to precipitate vitriolated magnefia, or muriated mag-nefia, by volatile alkali, he will indeed obtain fome precipitate, but a new triple combination, a falt of a peculiar

nature

nature, will be formed. If faturated folutions of nitrated lime and nitrated
magnefia be mixed, an unexpected precipitate appears, confifting of a triple
falt, compounded of both earths and the
common acid, more difficult of folution
than either of the ingredients, and on
this account falling to the bottom. The
new falt is taken up by a larger quantity of water. I muft overpafs many
phænomena of this nature.

IX.

*Exceptions from a determinate Excefs of one
or other of the Ingredients.*

SOME chemifts, I know, contend, that
it is idle to fuppofe that a determinate
excefs of acid can be received by neutral or middle falts. Many inftances,
however, which I fhall now mention,
clearly prove the prefence of fuch excefs,

cefs, which, agreeably to the nature of the thing, adheres more loofely than the faturating portion. Let perfectly neutral tartarized tartar be diffolved to faturation in diftilled water; then let fome genuine acid of tartar (XXIII.) be dropped in, and a v hite fpongy fubftance will feparate and fall to the bottom, which, when collected and examined, proves to be real tartar. What is the caufe of this fingular alteration? We fhall eafily afcertain it by confidering the nature of the fubftances. Tartar is nothing but vegetable alkali with a greater portion of its own acid than is neceffary to faturation. He who is acquainted with the tafte of tartar, its effervefcence with alkalis, the red colour it gives to blue vegetable juices, &c. can entertain no doubt concerning the excefs of acid; nay, even till our own times, tartar was confidered as an acid. Take away the abundant acid by the addition of vegetable alkali, and you

will

will have tartarized tartar, which the French call vegetable falt, *fel vegetal.* Purified tartar is therefore nothing but tartarized tartar with a determinate excefs of acid ; and when this is added to tartarized tartar, it is immediately generated, and, for want of a fufficient quantity of water to diffolve it, falls in great meafure to the bottom. Tartar, therefore, and tartarized tartar, differ not in the nature, but the proportion of their ingredients ; neverthelefs this caufe produces a wonderful difference in tafte and other properties, and efpecially in folubility. For tartarized tartar attracts water fo forcibly, that it commonly deliquefces in moift air : on the other hand, one part of tartar requires 150 parts for its folution in a middle temperature ; which is fo much the more furprifing, as we are certain, that the fuperfluous acid, by itfelf, as well as tartarized vegetable alkali, readily unites with water. The excefs, which occa-

fions

fions the difference, can neither be removed by cryftallization, a moiftened
filter, or, in fhort, by any other way
but faturation.

WE have therefore a manifeft example, from which we may conclude,
that vegetable alkali, though faturated,
does not reject, but, on the contrary,
eafily admits an excefs of the acid of
tartar. There is here too a clear inftance of attraction between a neutral
falt and an acid of the fame fpecies as
that which enters into the compound. If
any other acid be poured into tartarized
tartar, tartar is alfo feparated; a phænomenon ufually explained by faying,
that the acid employed expels the tartar by fuperior attractive force. But
tartar is not a pure acid, as was long
fuppofed. Why then is the alkali united
to it expelled at the fame time? If the
precipitation arife from fuperior attraction, why fhould acid of tartar effect
it?

it? Why fhould vinegar, an acid really
weaker, (XXXVII.)? That we may diftinctly perceive what happens in this
operation, let the tartarized tartar be
imagined to be divided into two parts,
fo that one part *b* fhall contain as much
acid as is neceffary for the other *a* to
become tartar. Now let the foreign acid
be added, fo as to faturate the alkaline
bafis of the part *b*, the acid of tartar
before combined with it will flow back
to the portion *a*, which already tends
to it with fo much force, that it immediately feizes it, and is converted into
tartar, provided any thing capable of
weakening the cohefion of the principles in *b* but in a fmall degree be
added.

SALT of Seignette fhews the fame
phænomena. If a folution of volatile
alkali be gradually faturated with acid
of tartar, another fpecies of foluble tartar will be formed, which is immediately

ly clouded by excefs of acid, and a new tartar is exhibited, very difficultly fo-luble, but, on account of the loofer connection of the principles, more acid than the common fort.

But it is not only tartar which ef-fentially requires an excefs of acid. We have long been acquainted with feveral falts of this kind. Salt of forrel confifts of vegetable alkali and a peculiar acid in excefs, (XXIV.). So alfo acid of arfe-nic, precifely faturated with vegetable alkali, cannot be cryftallized ; but if there be a proper excefs of acid, we eafily obtain beautiful cryftals, (XX.). Hence it appears why it has hitherto been impoffible to prepare Mr Mac-quer's arfenical falt in a crucible; for the neceffary excefs has always been expelled by the force of fire.

Duhamel and Grosse have obfer-ved, that foluble tartar may be prepared

red from the abforbent earths, without
however underftanding the real nature
of the operation. Now, as upon the
addition of alkali, the excefs of acid is
faturated, and the whole mafs becomes
foluble, fo chalk, by abforbing this excefs, immediately generates a falt of
difficult folubility, which of courfe is
precipitated, (XXIII.); but when the
excefs of acid is feparated from the tartar, nothing but tartarized tartar, which
is very foluble, remains.

In 1760, Mr Baumé publifhed an experiment highly deferving of attention,
from which he thinks it evident, that
vitriolated tartar may be totally decompofed by nitrous acid in the humid
way. By this inftance, in the opinion
of fome modern writers, reciprocal affinities are proved beyond all doubt;
but a clofer examination will diffipate
the whole ambiguity. It is therefore
to be obferved, 1*ft*, That vitriolated tar-

tar,

tar, diffolved in water, may be cryftallized by evaporation, after the addition
of a quantity of concentrated vitriolic
acid, equal to one third of the falt. The
cryftals, with the acceffion of one third
of their weight, remain dry, notwithftanding they are acid. More acid affords a deliquefcent falt. The excefs
of acid cannot eafily be driven off by
diftillation in a retort; this end may
be more readily obtained by fufion
in a crucible. Repeated cryftalliza-
tions are of no avail. Wafhing with
highly rectified fpirit of wine is the
beft method of edulcoration. 2*dly*,
We know, that vitriolic acid in proper
quantity completely decompofes nitre
even in the moift way, whence its fuperior power of attraction is evident.
There is here, therefore, no occafion
for a diftinction betwcen the dry and
the moift way. 3*dly*, A third part only, or a very little more, of vitriolated
tartar, diffolved in ftrong and hot nitrous acid, is decompofed, whatever
quantity

quantity of the acid be employed.
4*thly*, There is no occasion to apply
heat, or use concentrated nitrous acid ;
for to a portion so much diluted that it
emitted no fumes, I added a large quantity of powdered vitriolated tartar, set
it in a cool place for thirty-six hours,
and then poured off the liquor ; from
which highly rectified spirit of wine
precipitated a white powder, which being collected and dried, proved to be
real nitre ; and it deserves to be remarked, that the vitriolated tartar which was
not decompofed, was so foluble by the
aid of the superfluous acid, as to be
scarce feparable by spirit of wine.
5*thly*, Vitriolated tartar, with a proper
excess of acid, as that in obferv. 1. is
not at all changed by the moft concentrated nitrous acid. It is scarce sufficient to moiften the vitriolated tartar
in powder with vitriolic acid ; they
muft be diffolved together in hot water.
6*thly*, Not only the nitrous, but the marine, the tartareous, and perhaps many
other

other acids, in like manner decompose vitriolated tartar. Glauber's falt, or vitriolated mineral alkali, is alfo totally foluble in marine acid; but about a third part only is decompofed, as Mr Kirwan has obferved. 7thly, Two thirds of the vitriolated tartar, which remain unchanged, form cryftals with the excefs of vitriolic acid, of the fame nature with thofe which are procured in the way mentioned in the firft of thefe confiderations.

If we weigh thefe obfervations, it will plainly appear that the fame thing happens in the prefent cafe, as in that of tartarized tartar. Suppofe *b* to be fuch a portion of the vitriolated tartar, as to contain exactly that excefs, which the other portion *a* can receive. Nitrous acid of itfelf cannot deprive the vitriolic of its bafis; but *a* attracting it at the fame time, fo far diminifhes the refiftance, that the nitrous is able

to

to feize the alkaline bafis of *b*, but its power is confined to certain limits. Suppofe the vitriolated tartar to be divided into two parts, one of which affords its bafis to the nitrous acid, and the other is not decompofed. We have here three powers : let that by which the part of the vitriolated tartar remaining entire attracts a determinate excefs of acid, be called *A* ; *B*, that by which the part to be decompofed endeavours to retain its bafis ; and, laftly, *C* the force of attraction of the nitrous acid to the fame bafis, it is obvious that no decompofition can be effected, if $A + C < B$, or if $A + C = B$; but if $A + C > B$, it immediately takes place.

WHAT has been faid concerning the folution of vitriolated tartar in nitrous acid, is in like manner applicable to Glauber's falt, fecret fal ammoniac, and perhaps many others, fo that thofe decompofitions cannot be deduced from
the

the prefence of phlogifton, in the alkaline falt. Concentrated folutions of
nitre and digeftive falt yield, upon the
addition of acid of tartar, a real tartar,
for the reafons above affigned; but
quadrangular nitre and fea falt, of
which the bafis, mineral alkali, has a
far different attraction for acid of tartar, afford no precipitation in experiments of this kind.

SEVERAL apparent exceptions originate from the removal or diminution
of excefs of acid; for various fubftances produce, with certain menftruums, falts fo difficult of folution,
that they cannot be held fufpended
without fome excefs. Thus lime is foluble in abundant acid of arfenic; but
cauftic volatile alkali, magnefia, lime itfelf, and, in fhort, whatever is capable
of abforbing the abundant acid, immediately produces precipitation. If any
one fhould hence conclude that lime is
expelled

expelled by cauſtic volatile alkali and
magneſia, he is certainly deceived, and
ought likewiſe to maintain, that this is
done by the lime itſelf. The precipitate, when examined, does not exhibit lime alone, but lime ſaturated
with arſenical acid, which ſufficiently
explains the nature of the operation.
The ſame phænomena occur with lime
diſſolved in phoſphoric acid, and with
many other ſubſtances of difficult ſolubility.

Almoſt all the metallic ſalts redden
tinſture of turnſole; and the exceſs
can ſcarce be removed without deſtroy-
ing the ſalt.

But it is not the acids alone which
ſometimes exceed the limits of ſaturation; this is likewiſe true of the ſaline,
earthy and metallic baſis. Borax, however well purified, exhibits clear marks
of abundant alkali, and ſtill requires
about

about an equal weight of fedative falt, to be completely faturated. Why the arfenical acid, though perfectly faturated with vegetable alkali, fhould yet expel the acid of nitre in diftillation, I have already fhewn, (VI.) ; but I may here add that the acid of arfenic like-wife attracts an excefs of alkali, when circumftances allow, and this force undoubtedly promotes the feparation. On the fame principles, the acid of arfenic, exactly faturated with vegetable alkali, decompofes liver of fulphur and foap, as Mr Scheele has difcovered. In alum there is an excefs of acid, fo that it reddens turnfole, and is capable of receiving a ftill greater excefs, and reciprocally of being combined with its own bafis beyond the bounds of faturation. The calx of lead may alfo be combined in excefs with plumbum corneum, and faccharum faturni. Turbith mineral and powder of algaroth have an excefs of their bafis, and, after

the

the moſt careful waſhing, yield, on diſtillation, a portion of acid. I omit
other inſtances; and from thoſe which
have been adduced, I think it evident
that the doctrine concerning a determinate exceſs of one or other of the
ingredients, is not only not abſurd, but
that it actually takes place on many oc-
caſions. The exceſs commonly adheres leſs firmly than the portion re-
quiſite for ſaturation, and therefore in
many inſtances may be eaſily removed,
but it is not on this account the leſs
real. There is in theſe caſes, as I have
before remarked, an attraction between
the ſaturated ſalt, and a determinate
exceſs of the acid or the baſis. Perhaps ſuch an attraction takes place in
all compound ſalts, and ſometimes the
power which attracts the acid, and at
others, that which attracts the baſis,
may prevail, though we are as yet acquainted with only a few inſtances. It
is alſo probable that the ſaline particles,
when

when diffolved, can admit of a greater
excefs than when in a concrete ftate;
at leaft, fuch is their relation to the
matter of heat, a fubftance far more
fubtile, for when they coalefce after
they have been feparated, they part
with a certain portion which they attract when diffolved. A new field
opens here before us, as yet unculti-
vated, and indeed fufficiently difficult,
fince the attraction of compounds is
weaker and fometimes fcarce perceptible; fometimes, however, remarkable
phænomena are to be derived from
them alone. Let mercury, for inftance,
be digefted in an equal weight of nitrous acid, with fuch a degree of heat
as will prevent cryftallization. At
firft the metal is taken up with effervefcence in the common manner, but
at length the generation of bubbles
ceafes, nor does any nitrous air arife,
though in the mean time moft of
the mercury infenfibly difappears. In
this

this experiment, ordinary nitrated mercury, with a calcined baſis, is formed, and this is afterwards ſaturated with mercury, that retains its phlogiſton. If a ſolution of ſea ſalt be added to a ſolution of this ſalt, a white powder is precipitated, which is real mercurius dulcis, and which in the laſt Swediſh Pharmacopœia is directed to be prepared in this way*. When the mixture is made, the marine acid attracts the calcined mercury, and forms corroſive ſublimate, which immediately ſeizing the complete mercury, becomes perfectly mild; nor does any thing elſe happen when calomel is prepared in the dry way.

X.

---

* Scheele in the Stockh. Tranſactions, 1778.

X.

How we are to determine the single Elective
Attractions.

AFTER this view of the difficulties
which may occur, let us haften to our
fubject. Suppofe *a, b, c, d,* &c. to be
different fubftances, of which the attractive forces for *A* are to be afcertained.

A.] Let *A d,* (i. e. *A* faturated with
d,) be diffolved in diftilled water, and
then add a fmall quantity of *c,* which
may either be foluble in water by itfelf
or not. Firft let it be foluble ; then a
concentrated folution ought to be employed, which, when dropped into a folution of *A d,* fometimes immediately
affords a precipitate, which, being collected and wafhed, either proves to be
a new combination, *A c,* with peculiar
properties,

properties, or *d* is extruded, or fometimes both. It now remains to be examined, whether the whole of *d* can be
diflodged by a fufficient quantity of *c*
from its former union. It fhould be
carefully noted in general, that there is
occafion for twice, thrice, nay fometimes fix times the quantity of the de-
component *c*, than is neceffary for faturating *A* when uncombined. If *c* effect
no feparation, not even in feveral hours,
let the liquor ftand to cryftallize, or at
leaft become dry by a fpontaneous evaporation ; high degrees of heat muft be
avoided, left they difturb the affinities,
(IV.). Here the knowledge of the
form, tafte, folubility, tendency to ef-
florefce, and other properties, even thofe
which, in other refpects, appear of no
confequence, of the fubftances, is of
great ufe in enabling us to judge fafely
and readily, whether any, and what decompofition has taken place. Sometimes the difengaged fubftance, whether
that

that which was added or expelled, gives
the operator much trouble, by conceal-
ing the genuine properties of the other,
and therefore, if poffible, fhould be removed, according to circumftances,
either by water or fpirit of wine.

NEXT, fuppofe *c* to be infoluble, as,
for inftance, a metal, let a bright and
clean plate of it be put into the folution
of *A d*, and let it be obferved, whether
any thing is precipitated. By putting
feveral laminæ in fucceffion, we find at
laft whether a part only of *d*, or the
whole, is feparated. Sometimes no decompofition is effected, though the furface of the metal fhould have been lately filed, unlefs there be a fmall excefs
of acid ; and as far as I have hitherto
been able to collect, it is not always
of confequence that the fuperfluous acid
fhould be of the fame nature as that
which *A d* contains or not.

IF

IF only one of the compounds *Ad* and *Ac* be foluble in highly rectified fpirit of wine, there is fcarce any need of evaporation ; for if the mixture be made, and left a few hours at reft, and then fpirit of wine be added, that which cannot be diffolved in it is feparated.

THE fmell alfo often indicates what is taking place. Thus, vinegar, acid of ants, of falt, nitre, volatile alkali, are eafily diftinguifhed when fet free. The tafte likewife often informs an experienced tongue.

b.] Let *Ad* then be treated with *b* and *a*, &c. feparately in the fame manner.

c.] In like manner, let *Ac, Ab, Aa,* be examined in their order.

By

By fuch an examination properly conducted, the order of attractions is difcovered. This tafk, however, exer-cifes all the patience, and diligence, and accuracy, and knowledge, and experi-ence of the chemift. Let us fuppofe only a feries of five terms, *a, b, c, d*, and *e*, to be examined with refpect to *A*, twenty different experiments are requi-fite, of which each involves feveral others : a feries of ten terms requires ninety experiments, and, in general, if *a* be the number of the feries, $n. \overline{n-1}$ will be the number of experiments.

d.] In like manner, each compound with *a, c, b,* fhould be examined in the dry way ; but it muft be in a crucible, or, if poffible, in a retort heated to incandefcence, that the volatile part may be collected at the fame time.

Such, in general, is the method which I have followed ; the continuance of
this

this labour will perhaps difcover various fhorter paths, which will at leaft be
convenient in certain eafes. But we
fhould be cautious in guarding againft
fallacies arifing from the apparent exceptions above defcribed.

XI.

The neceffity for a new Table of Attractions.

THE tables which we have at prefent contain only a few fubftances, and
each of thefe compared only with a few
others. This is no reproach to the authors of them, for the tafk is laborious
and long. Although, therefore, I have
been employed upon it with all the diligence I could exert, and as much as
my many other engagements would permit, yet I am very far from venturing
to affert, that that which I offer is perfect, fince I know with certainty, that
the

the flight fketch now propofed will require above 30,000 exact experiments,
before it can be brought to any degree
of perfection. But when I reflected
on the fhortnefs of life, and the infta-
bility of health, I refolved to publifh
my obfervations, however defective,
left they fhould perifh with my papers,
and I fhall relate them as briefly as poffible. In itfelf it is of fmall confequence by whom fcience is enriched ;
whether the truths belonging to it are
difcovered by me or by another. Meanwhile, if God fhall grant me life, health,
and the neceffary leifure, I will perfe-
vere in the tafk which I have begun.
I fhall now explain the end I had in
view, and my plan ; fhould they be ap-
proved by the mafters of the fcience, I
hope that many will lend me their
affiftance, for it is eafier to accomplifh
one or two columns, than to bring all
to perfection : I exhibit a great number of the more fimple fubftances which

occur

occur in chemiftry. Many of thefe are
not only compounded, but are eafily refolved into their proximate principles,
fuch as hepar, fulphur, the imperfect
metals, &c. ; but they do not come into view here, but inafmuch as they effect compofition and decompofition in
their entire ftate ; but when their proximate conftituent parts are feparated,
double attractions take place, which are
not confidered in this table.

MOREOVER, I have inferted many
lately difcovered, of uncertain origin
and compofition, fuch as the acids of
fluor, arfenic, tartar, fugar, and forrel ; of
earths, magnefia and terra ponderofa ;
of metals, platina, nickle, manganefe,
and fiderite, of which more in the place
belonging to each. In the obfcurity of
their origin, thefe fubftances agree with
others that have been the longeft known.
Should they be derived from others,
they ought not, on this account, to be
excluded,

excluded, for they are now different, have conſtant properties, exerciſe their attractive powers without decompoſition, and can at pleaſure be obtained perfectly alike. It is therefore proper to inquire into their powers. Every ſubſtance that we employ is probably compounded, and although we are at preſent ignorant of its principles, they may hereafter be detected.

THE upper ſtratum of the table, if I may ſo call it, contains fifty-nine rectangles horizontally placed, which exhibit fifty-nine different ſubſtances, denoted by ſigns formerly in uſe, or by new ones, which I ſhall now therefore enumerate in the order of the adjacent numbers, for there is ſcarce any one in the following which does not appear in the firſt : 1. Is vitriolic acid ; 2. Phlogiſticated vitriolic acid ; 3. Nitrous acid ; 4. Phlogiſticated nitrous acid ; 5. Muriatic acid ; 6. Dephlogiſticated muriatic

muriatic acid; 7. Aqua regia; 8. Fluor
acid; 9. Arſenical acid; 10. Acid of
borax; 11. Acid of ſugar; 12. Acid of
tartar; 13. Acid of ſorrel; 14. Acid of
lemon; 15. Acid of benzoin; 16. Acid
of amber; 17. Acid of ſugar of milk;
18. Diſtilled vinegar; 19. Acid of milk;
20. Acid of ants; 21. Acid of fat;
22. Acid of phoſphorus; 23. Acidum
perlatum; 24. Acid of Pruſſian blue;
25. Aerial acid; 26. Pure fixed vegetable alkali; 27. Pure fixed mineral alkali; 28. Pure volatile alkali; 29. Pure
ponderous earth; 30. Pure lime;
31. Pure magneſia; 32. Pure clay;
33. Pure ſiliceous earth; 34. Water;
35. Vital air; 36. Phlogiſton; 37. Matter of heat; 38. Sulphur; 39. Saline
liver of ſulphur; 40. Alcohol; 41. Æ-
ther; 42. Eſſential oil; 43. Unctuous
oil; 44. Gold; 45. Platina; 46. Silver; 47. Mercury; 48. Lead; 49. Copper; 50. Iron; 51. Tin; 52. Biſmuth;
53. Nickle; 54. Arſenic; 55. Cobalt;
56.

56. Zinc; 57. Antimony; 58. Manganese; and, 59. Siderite.

THESE fubftances are, as it were, the heads of each column, at the top of which they refpectively ftand : to thefe thofe that are placed below bear this relation, that the nearer they ftand, the ftronger attraction they muft be underftood to have. Every column, therefore, not only muft exhibit every one of the fifty-nine fubftances which is capable of being combined with the principal fubftance at the top, but alfo the order which fuch combinations follow. The double line diftinguifhes from the others the thirtieth ftratum, which is the firft that belongs to the dry way. The fubftances which occur in thefe rows refer alfo to the heads of the columns.

LASTLY, I have diftinguifhed the horizontal rows, as well as the columns, by numbers on each fide, that each

rectangle

rectangle might be more readily found
and quoted. On account of the new
fubftances, I am obliged to divide the
table of fingle elective attractions into
two parts; and when, from multiplied
experiments, more than two can contain
fhall require admiffion, it may be conve-
niently divided into four parts; the firft
for the acids, the fecond for the alkalis
and earths, the third for the inflam-
mables, and the fourth for the metals.

XII.

Column Firft, the Vitriolic Acid.

CONCERNING the head of this column, as it is fo well known, there is no
occafion to premife much. So firm is
its compofition, that its proximate principles have not yet been difcovered.
Some late excellent experiments * have
been

* Lavoifier's, in the Mem. of the Acad. of Paris, 1777.

been thought to difclofe the ftructure
of this acid; but they are, if I miftake
not, to be underftood in a different
manner. Sulphur, when burned in a
veffel filled with atmofpheric air, and
clofed by means of mercury, abforbs
a portion of vital air, and yields an acid
of twice or thrice the weight of the
burned fulphur. The acid, therefore, is
fuppofed to have exifted in fulphur, far
lighter, and without air. The fame thing
is confirmed by the efflorefcence of ful-
phureous pyrites, which is converted
into vitriolated iron, not however without the abforption of a certain portion
of vital air. To recover the air inherent in this acid, vitriolated mercury is
reduced in a pneumatic apparatus by
the aid of fire to its metallic form. During this operation, a large quantity of
vital air, which is fuppofed to enter into
the compofition of the acid, is collected. Of thefe facts the following feems
the true explanation. It has been fully
                                    proved

proved by experiments, that every fubftance has a certain fpecific quantity of
fire, which yet varies more or lefs in
one and the fame, according to the different ftates of folidity, liquidity, and
fluidity, (XLVIII.). Now the vitriolic
acid exifts in a folid ftate in fulphur,
but, on deflagration, deliquefces, and
therefore recovers the heat proper to its
liquid ftate. The fpecific heat of fulphur is to that of vitriolic acid as o, 183
to o, 758, that is, nearly as 1 : 4. But
the acid extricated in this experiment
contains very little water, only the
quantity indifpenfably neceffary to fluidity, which it attracts from the air and
the mercury, that almoft always contain it. But the lefs water the acid
contains, the lefs is its fpecific heat,
and it undoubtedly, in this cafe, is below o, 758 : let us fuppofe it to be
o, 549, and thus the proportion will be
changed to 1 : 3. It follows, then, that the
vital air enters into the compofition of the
fpecific

fpecific fire.   It is evident, that it lofes
its aerial form during the combination,
which it cannot regain without a decompofition, fince, in other experiments, it
may be expelled from acids by alkalis
and other faturating fubftances, in the
fame manner as the aerial acid out of
chalk; but the heat only is fet at liberty, and no part of the vital air.   That
which appears in the reduction of vitriolated mercury perhaps arifes from
decompofed heat, as we fhall more
clearly fee in XLVIII.   The principles
of vitriolic acid have not yet therefore
been fatisfactorily fet loofe ; for as to the
matter of heat, it exifts in every body
yet known.

Among the fubftances hitherto tried,
vitriolic acid adheres moft tenacioufly
to

2.] *Cauftic terra ponderofa*, which, when
added to a folution of vitriolated tartar,
generates

generates the ponderous fpar, which remains infoluble at the bottom. The liquor contains cauftic vegetable alkali, (Scheme 1.). Cauftic or pure vegetable alkali is incapable of decompofing ponderous fpar.

3.] Next ftands *cauftic vegetable alkali*, which, when added in fufficient quantity to a folution of Glauber's falt, yields vitriolated tartar and uncombined mineral alkali, which is unable to detach the vitriolic acid from vegetable alkali.

4.] Cauftic *mineral alkali* precipitates the calcareous bafis of gypfum, but the inverfe experiment does not fucceed.

5.] *Cauftic calcareous earth* is fuperior to magnefia; for vitriolated magnefia (Epfom falt) is immediately decompofed in lime-water, and yields its acid to the lime. Moreover, lime feparates volatile
alkali

alkali and all the metals from vitriolic
acid.

6.] *Cauftic magnefia* added to a folution of fecret fal ammoniac, feems to
produce no change which is fenfible to
the fmell; but if the mixture be kept
for a few days in a clofe phial, a diftinct
fmell of volatile alkali will be perceived
on opening it. The difference however of force is very fmall; fo that the
fmalleft diminution of that of the former, or increafe of that of the latter,
inverts the attractions. Hence a precipitation of vitriolated magnefia is often
effected by cauftic volatile alkali: for
the alkali cannot eafily be obtained
quite pure, being either, on the one
hand, contaminated by a fmall quantity
of aerial acid, or, on the other, by
quicklime, either of which effects a feparation, the former by means of a
double (V.), the latter by a fingle attraction. But the chief and perpetual
caufe

caufe of precipitation, is the formation
of a triple falt, more difficult of folution,
as we have before explained, (VIII.).

7.] *Cauftic volatile alkali* precipitates
clay from vitriolic acid, and zinc likewife, unlefs it be added in fufficient
quantity to rediffolve the precipitate,
(VII.). The fame caution is applicable
to the other metals ; but there will be
no ambiguity in the refult, if the metal
be infoluble in the precipitate. See alfo
what is faid of volatile alkali in XVI.
and XXXIX.

8.] *Pure clay*, i. e. earth of alum, long
digefted in alkaline water, and then
well edulcorated. I have already treated at fufficient length concerning the
precipitation of alum by zinc, iron, and
fome other metals, (VI.). But the metallic calces feem to have the fame degree of attraction for acids as clay, at
leaft I have in vain tried to decompofe
vitrio-

vitriol of copper by clay ; and reciprocally calx of copper tinges a folution
of alum, and a white fediment is depofited ; but this effect, as I have before
explained, is owing to excefs of acid.

9.—23.] This fpace perhaps belongs
to the *metallic calces.* In all the tables
of attractions which have been publifhed, and even in that which I offered
to the world in 1775, the metals were
placed in the columns of the acids ;
but upon farther reflection, I am forced
to exclude them. That thefe fubftances are attracted and diffolved by
acids, is known even to beginners ; but
let it be remembered that they are not,
as was fuppofed, taken up entire, and
in their complete form by menftrua :
for fome particles of the acid carry off
the fuperfluous phlogifton, while others
diffolve the calcined metal. Since
therefore they exift in the menftruum
mutilated, and in a great meafure deprived

prived of one of their principles, the condition under which the procefs may be
referred to fingle attractions, does not
exift. Hitherto the precipitations of metals by metals have been ill underftood.
When I obferved many years ago, for the
firft time, that the feries of the metals
was the fame with refpect to all the
acids, I was ftruck with great furprife at
the coincidence, confidering in how many particulars earths and alkalis differ
with regard to them. I therefore began to entertain a fufpicion, that the
precipitation of metals depended, not
on the election of the acids, but on
fome other principle, which I now certainly know to be the attractive power
of the diffolved calces for the phlogifton of the precipitating metal. I
have elfewhere treated of this fubject *,
and fhall fay more upon it in the
fequel, (XLVII.). Complete metals
therefore are properly excluded, but
are

* Diff. de phlogifti quantitate in div. metallis, § 2.

are the calces alfo to be fet afide? They are really diffolved; and it feems agreeable to the nature of things to fuppofe, that the fame acid would find fome difference in fixteen different calces, in confequence of which it would prefer fome to others. But as reafoning is fallacious, without the teftimony of experience, I performed experiments with the calces, efpecially with thofe of filver and copper. I firft procured as faturated a folution of filver in the nitrous acid as poffible, which I could not indeed bring to fuch exactnefs that it would not redden turnfole; but the excefs can fcarce be taken away without the precipitation of the metallic falt. To this folution I added copper calcined by fire, and expofed it to a heat of digeftion for feveral days; but though it was only very flowly diffolved, and the colour of the liquor changed to a blue, no figns of precipitation appeared. Another folution of filver,

filver, equally faturated, diffolved that calx of mercury which is ufually called precipitate *per fe*, without any diminution of its tranfparency. I afterwards faturated nitrous acid with copper, and added the calx of filver precipitated by cauftic fixed alkali ; here there was only an inconfiderable folution, and no precipitation at all. It feems, therefore, that an acid takes up calcined metals without diftinction, provided they have loft a certain quantity of phlogifton ; for more or lefs of this principle makes a remarkable difference in fome cafes. Befides, when the nitrous acid is ufed, it fometimes happens that a calx, in a proper ftate, is gradually deprived of its phlogifton beyond the determinate limit, and then it is immediately rejected. Such events, proceeding from the peculiar nature of certain fubftances, muft be carefully obferved, left erroneous conclufions fhould be drawn. I have before obferved,

ferved, that the calces attract each o-
ther, particularly thofe of zinc and copper *. Thefe combinations, when diffolved in the fame acid, produce, without doubt, triple falts, which deferve
farther examination.

I have inferted the metallic calx in
the order in which they are ufually
precipitated, fince it may not be without ufe to be acquainted with it ; but
I have obliterated the horizontal lines,
in order to fhew that the acids have
not yet been found to poffefs any power
of felection.

24.] I here place *water*, fince it diffolves moft of the vitriols, and reftores
them unchanged. I am aware, indeed,
that mercury, tin, bifmuth, and antimony, are feparated from the vitriolic
acid, upon the addition of water ; but
it fhould be obferved, at the fame time,
that

* *Ibid.* § 5. d.

that a large quantity is requifite, which firft carries off the excefs of acid efpe-cially neceffary to thefe falts, and more flightly adhering to them, and then by the concurrence of heat feizes the re-mainder ; but a proper quantity does not render the folutions turbid. We know that the vitriolic acid cannot be perfectly deprived of its fuperfluous water by boiling, but that it retains more than one-fifth of its weight; which therefore is the leaft poffible quantity in metallic folutions. But by a fuffici-ent quantity of water, and in a certain length of time, all the vitriols perhaps may be decompofed; in which cafe ano-ther place fhould be affigned to it, un-lefs fome other caufe exerts its influ-ence at the fame time.

25.] *Phlogifton* comes laft, to which, however, fome of the moderns give the firft place ; but I am as yet unacquaint-ed with any experiment from which

it

it can be safely concluded that phlogiston, in the humid way, and by attracting the acid, is capable of decomposing either neutral or middle salts, whether earthy or metallic. It is indeed strongly attracted by the vitriolic acid, as appears from the dark colour which it contracts from the smallest portion of oily matter, whether this be uncombined or intimately united with some other substance ; however, a sufficient quantity of water both prevents the offuscation, and removes it when it has been long present. Moreover, this acid, though in the most concentrated state, does not affect the phlogiston of charcoal, except by means of a proper degree of heat. Metals put into the vitriolic acid, lose a certain portion of phlogiston, but this is the effect of heat ; at least, of that degree which is excited by the solution ; and I have before observed that this privation is necessary ceffary to folution, (V.) and fhall bring farther proof in XIII.

31.] In the dry way, *phlogifton* oc-cupies the firft place, for vitriolated tartar, Glauber's falt, ponderous fpar, and gypfum, lofe their acid by the in-tervention of the inflammable principle of charcoal, and a fufficient heat.

32.] It is probable, that *terra ponde-rofa* decompofes vitriolated tartar in this way; it remains, however, to be confirmed by experiment.

33.] *Vegetable alkali* expels the vo-latile.

34.] So does *mineral · alkali* ; but whether the latter yields to the vege-table has not yet been examined.

35.] *Lime,* as well as

36.]

36.] *Caustic magnesia* deprives secret sal ammoniac of its acid.

37.] All the *metals* probably, or rather their calces, expel the cauftic volatile alkali. Experiments have been made with lead, tin, copper, iron, *&c.*

38.] *Volatile alkali.*

39.] *Pure clay* cannot detach the acids from ammoniacal salts.

XIII.

Column Second, the Phlogisticated Vitriolic Acid.

It was an emphatical and juft obfervation of the ancients, that phlogifton lent wings to vitriolic acid, which, though it requires an intenfe heat to be fublimed before its union with phlogifton,

gifton, afterwards evaporates fpontaneoufly. The effects, however, vary
wonderfully, according to the different
proportions. For the acid, when fully
faturated, conftitutes common fulphur;
if it be combined with a fmaller quantity, it generates aeriform vitriolic a-
cid, known likewife by the name of vitriolic acid air, which, when collected
in mercury, cannot be condenfed into a liquid by cold; is very light, not exceeding
000, 246 in fpecific gravity. It immediately diffolves camphor, and extinguifh-
es flame. An hundred grains of diftilled water, fcarce take up five of
this aeriform acid; and I call this liquor, for the fake of diftinction, *phlogifticated vitriolic acid.* This acid freezes
in the fame temperature as pure water;
and what is remarkable, the acid elaftic fluid remains in the ice, though in
open veffels it forfakes the water.
What is highly worthy of notice, is,
that if it be expofed to heat in a tube
hermetically

hermetically fealed for twenty days, a fmall quantity of fulphur is feparated. Has the decompofition of heat any fhare in this phænomenon?

THE vitriolic acid, by the aid of fire, and properly treated, may be phlogifti-cated by moft fubftances containing the inflammable principle; but it cannot be reduced to this ftate by means of aerial acid. The phlogifton, in thefe operations, works wonderful changes, for a very fixed, heavy, inodorous, a-crid liquor, becomes elaftic, light, and fo volatile, that its very penetrating fmell threatens fuffocation, and more-over fo weak, that vegetable acid at-tracts alkali from it. I have not yet learned from experiment, whether there hence arifes any variation in, the elec-tive attractions. I know that it dif-folves alkalis; that cauftic fixed alkali, and pure lime expel volatile alkali; and alfo, that lime-water precipitates
magnefia;

magnefia ; and till the order of the o-
ther fubftances fhall have been afcertained, I follow the fame as in the preceding column. The neutral and
middle earthy falts, formed by this a-
cid, differ a little in figure, tafte, and
other properties, from thofe which contain pure vitriolic acid ; the difference,
however, difappears in time, for the
phlogifton gradually flies off.

But as all metals, in order to be foluble, muft be deprived of a determinate portion of phlogifton, which, for
each, is various, (for none can be taken
up in its reguline ftate by vitriolic acid,
without the feparation either of inflammable air, or aeriform acid ; while, on
the contrary, each, when deprived of a
certain portion of phlogifton, is not only
more eafily diffolved without any farther
lofs of phlogifton, but likewife afford the
fame vitriols as in the preceding cafe ;)
hence it neceffarily follows, that the
phlogifticated

phlogifticated vitriolic acid fhould reject them; and there are fome experiments from which it would appear that
it really is fo. Zinc, which is quickly
diſſolved in diluted vitriolic acid, is
changed, by the fame acid properly
phlogifticated, into a white powder,
which feems neither to be taken up by
vitriolic nor marine acid. Each particle of the menſtruum muſt be loaded
with phlogiſton, otherwife thofe which
are free from it act at firſt in the uſual
way ; but when they are faturated, folution ceafes. It is faid, that, by the aid
of heat, the zinc is attacked, and that
a quantity of inflammable air is extricated ; but I have not yet feen this.
Flowers of zinc are taken up by
the phlogifticated acid. Iron agrees
with zinc, except that when it is too
much calcined, it is fcarce foluble.
Copper is not vifibly changed in this
menſtruum. Metallic precipitates procured by alkalis are by no means to be
                                    confidered

confidered as reguli minutely divided;
for they are more or lefs deprived of
phlogifton, as appears partly from what
has been faid above, and partly from
the fediments thrown down by any metal ufed as a precipitant ; which differ
from the former both in fplendour and
nature; I have therefore no doubt but
phlogifticated vitriolic acid will diffolve metals properly calcined ; but I
confefs, that the particular phænomena
have not been examined with proper
attention. This volatile menftruum
cannot be fubjected to experiments in
the dry way.

XIV.

Column Third, Nitrous Acid.

1.] This acid, likewife, feems to have
fuch firmnefs of ftructure, that its principles have not hitherto been afcertained.

ed*. Expofed to the fire with various
fubftances, it yields a great quantity of
vital air ; but it remains as yet doubtful whether the air exifted uncombined
in the acid, or is formed by its being fufficiently phlogifticated.   This queftion
will be difcuffed hereafter, (XLVI.).
It exerts its elective attractions nearly
in the fame order as the vitriolic acid.

2.] *Cauftic ponderous earth* cannot be
feparated from nitrous acid by the cauftic vegetable alkali.   When the aerial
acid is prefent, it is precipitated in
confequence of a double attraction; but
if there be too much, it will be rediffolved.

3.]

* The Count de Saluces politely fent me his letter on
the generation of nitre, but he has not yet publifhed his
proportions, and I have not therefore had the fatisfaction
of obferving this fpectacle.   Mr Thouvenel, too, has very
lately obtained the prize from the Parifian Academy,
for procuring nitrous acid from atmofpheric air and putrid vapour.

3.] *Cauſtic vegetable alkali* decompoſes quadrangular nitre, and forms priſmatic.

4.] *Cauſtic mineral alkali* precipitates
nitrated lime.

5.] *Lime* precipitates nitrated magneſia.

6.] *Cauſtic magneſia* expels the volatile alkali from nitrum flammans.

7.] *Cauſtic volatile alkali* precipitates
clay, zinc, and the reſt of the metals.

8.] The place of *pure clay*, as alſo of
the ſubſtances which follow has not
been ſufficiently determined.

9.—24.] The *metals* properly calcined.

25.]

25.] Water feems to prevent the acceffion of phlogifton. The acid juft expelled from nitre, by vitriolic acid, contains about two-thirds of its weight
of water.

26.] The nitrous acid foon detaches
from the metals that portion of phlogifton which impedes folution; and
when heat is employed, it fometimes
goes beyond proper bounds, infomuch,
that being too much calcined, they
cannot be held in folution. Thus, tin
and antimony are taken up with vehemence, but are foon let fall again to the
bottom.

In the dry way, the fame order as in
column firft, as far as I have yet been
able to learn, is obferved. Phlogifton
occupies the firft rectangle, for in detonation the acid forfakes both ponderous earth and vegetable alkali to u-
nite with phlogifton. Whether it be
converted into vital air, I do not here
enquire,

enquire, but the intenfity of the deflagration, even in vacuo, clearly fhews
the prefence of this air, which feems to
be totally converted into heat; for
nitre, in detonating with charcoal in
clofe veffels, yields fixed and foul air,
but fcarce any thing fit for fupporting
ignition and refpiration.

XV.

Column Fourth, the Phlogifticated Nitrous
Acid.

NITROUS acid, efpecially when concentrated, eagerly attracts the inflammable principle, and, when contaminated with it, emits red vapours, and the
liquor acquires a reddifh colour, which,
however, may be fo far driven off by a
flow diftillation, that the liquor fhall appear as limpid as the cleareft water;
fuch an acid is juftly called *pure.* But
the

the colour which has been made to difappear, returns upon the fmalleft addition of phlogifton ; even the folar rays
induce a yellow colour, and caufe the
acid to emit yellow fumes, as Mr
Scheele has obferved. Smoking nitrous acid, moreover, furnifhes a fine
proof, that different colours depend upon the different denfity of phlogifton ;
for if nitrous acid, when red and concentrated, be diluted with about a fourth
part of its bulk of water, it affumes a
beautiful green colour, and yet emits
red fumes ; but an equal, or a greater
portion of water makes it blue, while
twice or thrice its bulk deftroys all colour. The red fmoke which rifes fpontaneoufly, or may be driven off by heat,
preferves its elafticity in a clofe veffel,
and cannot be reduced to a liquid by
cold ; it is therefore properly called
aeriform nitrous acid. It is abforbed by
water, which, with a certain portion,
becomes blue ; with a larger, of a beautiful

tiful green ; when faturated, is yellow, and is then found to have received an increafe of one-third of its bulk. Such are the variations of *phlogifticated nitrous acid.* The blue fpontaneoufly emits *nitrous* air, the green fcarce any, and the yellow none at all. It well deferves to be noticed, that nitrous air fometimes exceeds the phlogifticated acid from which it has been expelled, tenfold in bulk, though water cannot receive above one-tenth *. Yellow nitrous acid, expofed to heat in a tube hermetically fealed, becomes of a more intenfe colour, the green or blue is turned yellow ; but by refrigeration the former hue is brought back. When the heat is continued for a long time, a colour permanent in the cold is acquired ; the tinging matter may be expelled in the form of a red vapour, and the acid will remain without colour ; but after refrigeration, the vapour again enters into the acid, unlefs

* Prieftley.

lefs it be changed by a long continued
heat *.

THE phlogifticated acid diffolves alkalis and metals, but it adheres to them
very loofely, (XXXVII.). A fufficient
quantity of acid, (unlefs it be quite faturated), in an open veffel, is not much
prevented by the phlogifton from diffolving metals, for the particles which are
contaminated, or which attract this volatile fubftance, fly off; and moreover,
this menftruum attacks them on account of their inflammable part, and
does not take up thofe which are calcined beyond certain limits. The calx
of manganefe, known alfo by the name
of magnefia nigra, furnifhes an admirable proof of the effects of a certain
portion of phlogifton ; for this calx
contains a very fmall quantity of the
inflammable principle, on which account a fcarce fenfible quantity is diffolved

* Prieftley.

folved by pure nitrous acid, unlefs upon
the addition of fugar, honey, or fome
other inflammable fubftance, capable of
affording the neceffary complement;
but the phlogifticated acid perfectly diffolves it. Thefe folutions precipitated
by alkalis afford a white powder, readily
foluble in acids, but which, by heat, is
turned black, and recovers the properties of magnefia nigra; the white fediment, therefore, is nothing but the calx
united with as much phlogifton as is
neceffary to its folution in pure acids;
but the regulus contains a fuperfluous
quantity, fince red vapours are formed
during its folution in nitrous acid.
Mercury diffolved in nitrous acid, in
the cold, depofits cryftals fpontaneoufly; by cauftic volatile alkali, it is
precipitated of a black colour, and differs in many other refpects from that
which, in confequence of the application of heat, has loft more of its phlogifton. The fame remark is applicable

to

to iron, and fome other metals.   Every
metal indeed mult be deprived of a portion of phlogifton; but if this procefs is
carried beyond certain limits, either no
folution takes place, or one widely different from a real folution.   The order of attractions, not having been fufficiently explored by experiment, is adjufted according to the preceding orders.

XVI.

Column Fifth, the Muriatic Acid.

1.] The acid of fea falt is nothing
but water more or lefs combined with
marine or muriatic air.   This air is
properly denominated *aeriform marine
acid*, of which diftilled water is capable
of abforbing one half of its weight, and
then yields *phlogifticated marine acid*.   The
acid recently expelled by vitriolic acid
from fea falt, commonly contains three-
fourths

fourths of water. This acid feems to exert its attraction in the fame way as the preceding, though indeed in fome cafes more obfcurely.

2.] *Ponderous earth,* diffolved in marine acid, cannot be expelled by pure vegetable alkali.

3.] *Pure vegetable alkali* expels the mineral. (See Scheme 3. and 32.).

4.] *Pure mineral alkali* expels lime, (Scheme 4.).

5.] *Pure lime* feparates magnefia, volatile alkali, and the metals.

6.] *Pure magnefia* is to be placed before volatile alkali, for the reafon before mentioned. The acid, magnefia, and volatile alkali, in a proper quantity, unite and form a triple falt: hence, in order to attain the proper proportion, a little

magnefia

magnefia is always feparated on the addition even of the pureft volatile alkali;
but it is only the quantity neceffary for
attaining this end.

7.] *Pure volatile alkali* has no power
againft lime, (Scheme 5.); but the aerated precipitates it in confequence of a
double elective attraction, (Scheme 36.).
It precipitates metals.

8.] *Clay.*

9.—24. *Metallic calces.*

25.] *Water.* (See XII.).

26.] We fhall fee in the next paragraph, the relation of *phlogifton* to marine acid. Some of the fubftances adduced in the preceding columns are
wanting in thofe which follow, for all
cannot be diffolved in each menftruum;
but I leave the rectangles which otherwife

wife would belong to them empty, that the difference may be the more ftri-king.

IN the dry way, it may be prefu-med, that the fame order is obferved as in 1. and 3. till experiments fhall have fhewn the contrary. I confidently pre-fume that the volatile metals, both with refpect to one another, and thofe which are fixed, have by no means the fame action as in the humid way. Corro-five fublimate is decompofed by all the acids, in confequence of a double elec-tive attraction, as was above explained with refpect to antimony; and this alfo is without doubt the cafe with lead, filver, and other metals faturated with acid of falt, which, when diftilled with antimony, yield butter of antimony. If we are unacquainted with this caufe, thofe experiments of Mr Pott, which fhew that corrofive fublimate yields a butter with regulus of arfenic, and that,

on

on the other hand, not a grain is obtained with white arſenic, are abſolutely unintelligible ; but, when we are acquainted with it, there remains no obſcurity in theſe phænomena.

XVII.

Column Sixth, the Dephlogiſticated Marine Acid.

THE illuſtrious Stahl reckons phlogiſton among the proximate principles of the nitrous acid. All the experiments which have been ſince made, ſhew, that this acid attracts the inflammable principle with great avidity ; but we cannot hence draw any concluſion with reſpect to its compoſition, unleſs we are to believe an axiom in moſt inſtances falſe and contrary to phænomena, according to which, thoſe ſubſtances which contain ſome common principles

of

of the fame nature have a greater attraction than thofe which are formed
of principles altogether different. Martial vitriol is not foluble in fpirit of
wine, though a dephlogifticated ley of
it is very readily foluble, not to mention a great number of other inftances.
No one, at leaft on this ground, fufpected the prefence of phlogifton in marine
acid, which fo obftinately rejected that
volatile principle ; but this is now certain, from the difcovery of the ingenious Mr Scheele * : for magnefia nigra,
which we have before confidered as almoft deftitute of phlogifton, attracts it
with fo much force as to decompofe the
marine acid in a heat of digeftion ;
it is perfectly foluble in this acid, and
is precipitated of a white colour from
it, which fhews the acceffion of phlogifton, (XV.). But the acid thus dephlogifticated, conftitutes an elaftic
fluid, of a light red colour, of the fame
smell,

* Stockh. Tranfactions, 1774.

ſmell, if the greater maſs be conſidered
as hot aqua regia ; it is not eaſily ſoluble in water, and ſcarce leaves an
acid taſte, when made to paſs through
it ; but if it be confined over water
for twelve hours, four-fifths are abſorb-
ed, and the reſiduum conſiſts of common air : that which has paſſed through
water is capable of making ſolutions,
but the unwaſhed is the moſt efficaci-
ous.   It ſhould therefore be collected
in cylindrical phials, ſucceſſively adapted to the neck of the retort, which,
when full, ſhould be cloſed with glaſs
ſtopples.   A little water is beforehand
put into the phials to abſorb the muriatic air.   Subſtances which are to be expoſed to it ſhould be put in with the
ſtopples.   It attacks phlogiſtic bodies
with great vehemence ; whitens all the
colours of vegetables ; reddens martial
vitriol ; diſſolves all the metals directly, and affords the ſame ſalts which are
formed by the acid entire, which may
                                              alſo

alfo be affirmed with refpect to earths
and alkalis ; it changes white arfenic
to a liquid acid, (XX.) ; always regain-
ing its original form when its lofs is reftored ; fo that this truth is fufficiently
proved both analytically and fynthetically. It fhould be carefully noticed,
that the red elaftic fluid is properly entitled to the name of dephlogifticated marine acid, and not the liquor
in the receiver, which, although it has
received a portion of the elaftic fluid,
yet confifts chiefly of common marine
acid, (XVIII.).

THAT the dephlogifticated acid
fhould form with alkalis, falts exactly
like thofe which contain the entire acid,
is a proof that they contain fome of the
inflammable principle by which the deficiency is fupplied. What I have feveral times before obferved, concerning the neceffity of depriving metals
of a certain portion of phlogifton, be-

fore they can be diſſolved in acids, is
admirably confirmed by the power poſſeſſed by the dephlogiſticated marine
acid, of diſſolving them all. This
ſeems to take place according as the
phlogiſton adheres more looſely to them;
but whether the order is the ſame as
that of the preceding acids, muſt be decided by future experiments. Its volatility prevents its action in the dry
way.

As marine acid is already ſufficiently
provided with phlogiſton, it refuſes a
larger portion in its liquid ſtate; but in
its aerial form, having a larger ſurface,
and being freed from its aqueous
cover, it ſeems to admit more; nay, even
to attract it with avidity, and when
ſufficiently ſupplied with it, to become
inflammable. It may perhaps be ſuſpected that dephlogiſticated marine
acid is nothing but the acid in an
aerial form; on compariſon, however,

I

I found a great difference, for the former is taken up by water flowly, not immediately; it does not become inflammable by decompofing phofphorus gradually, but attacks it inftantly, refolves it into white vapours, and regenerates aeriform marine acid; it melts neither ice nor camphor, effects no change either on nitre or alum, &c. That which is at firft collected, while a diftinct odour of aqua regia is perceived, from a mixture of muriatic acid, with half the quantity of magnefia nigra, by a gentle ebullition, contains about nine-tenths of common air; but that which is obtained towards the end, contains fcarce one-eighth. The foul air which was mixed with the dephlogifticated vapour, fuffers a fcarce fenfible diminution from nitrous air.

XVIII.

XVIII.

Column Seventh, Aqua Regia.

WE have now no difficulty in explaining why a mixture of nitrous and marine
acids ſhould be capable of diſſolving gold,
though neither of them of itſelf attacks
this metal.   Gold muſt firſt be deprived
of a portion of phlogiſton, after which
it is taken up by various menſtrua.
Now the nitrous acid, ſeizing phlogiſton
with great avidity, eaſily decompoſes
the marine, whether it be diſengaged
or united with any baſis, as appears
both from the ſmell of hot aqua regia,
which is exactly like that of the dephlogiſticated marine acid, and likewiſe from the effect, for this menſtruum, thus deprived of phlogiſton,
can repair its loſs from any metal ; in
conſequence of which, gold becomes
ſoluble, particularly in the marine
acid,

acid, (XVII.). Hence cryftals of gold, procured by fea falt diffolved in nitrous acid, (for the two acids uncombined fcarce afford any), and freed by edulcoration from the heterogeneous matters adhering to them, are found to contain the marine acid only. In this procefs, therefore, the nitrous acid has no other effect than to dephlogifticate the real folvent, as much as is neceffary : that acid, however, alone, when concentrated by long continued boiling, directly attacks the inflammable principle of gold, very fubtilely divided, fuch as occurs in the procefs of parting ; it then diffolves the calx, and retains it fo feebly, that it often falls down fpontaneoufly, or in confequence of fhaking. And this is the true explanation of Mr Brandt's experiment, who found that gold was foluble in nitrous acid *. There is no need to confider the other folutions by aqua

H 2   regia

* Acta Stockh. 1748.

regia one by one; we may only obferve, that this compound menftruum
does not always yield triple falts, for
in thofe cafes in which the nitrous or
the muriatic acids can of themfelves
effect a folution, the compounds commonly cryftallize feparately, at leaft
in part.

THE elective attractions here alfo
follow the order fet down in the preceding columns.

XIX.

Column Eighth, Fluor Acid.

How this acid may be expelled from
fluor by the vitriolic, has now been known
for a confiderable time *. When difengaged, it always affumes and retains
an aeriform ftate, till it comes in contact

* Stockh. Tranf. 1771.

tact with water, which abforbs it as
well as other aeriform acids; and we
thus procure *phlogifticated acid of fluor*,
which gradually corrodes glafs, extracting particularly the filiceous part.
This acid, however, in its aerial ftate,
acts upon glafs much more efficacioufly,
efpecially if the vapours be hot, which,
though they be loaded with filiceous
earth, conftitute a tranfparent elaftic
fluid. When it is diffolved in water, it
depofits part of the filiceous earth under the form of a white powder, but
the reft remains diffolved in the liquor.
Let me enter into a fhort difcuffion,
and enquire whether this be an acid different from every other, or mere muriatic acid, modified by an earthy bafis.
Peculiar properties diftinguifh it from
every other acid, from the vitriolic
and marine in particular, with refpect
to which doubts have arifen; for when
digefted with a little calx of filver,
and then depurated by gentle diftillation.

tion, it does not form argentum corneum; nor with fixed alkali does it
yield vitriolated tartar, or Glauber's
falt, or digeftive or fea falt; with lime
it regenerates fluor; with magnefia it
forms a cryftallizable falt; with terra
ponderofa, an efflorefcing compound,
and with clay, a fweet and vifcid falt,
like jelly: it alfo diffolves filiceous
earth itfelf, which totally rejects all
other acids, (XLIV.)

It is indeed true, that this acid is
generally adulterated with a little of the
marine, whence, without doubt, the re-
femblance of fmell; but is the origin
of the nitrous acid therefore to be deduced from the marine, becaufe both
are prefent in aqua regia? How fmall
the portion of the acid of falt is, appears from the very fparing precipitation of filver and mercury from the
nitrous acid. The fluor acid, as far as
I have yet found from experiment,
neither

neither derives its origin from the marine nor the vitriolic; at leaft I cannot comprehend how the intimate union
of an earthy bafis fhould produce fo
wide a difference. We know that
acids, in fuch a combination, become
in fome degree milder, and lofe their
acrimony. Why then fhould the fluor
acid, when refolved into vapours, cor-
rode, and fometimes perforate even
glafs, a property belonging to no other
yet difcovered, with whatever bafis it
may be united? If any one fhould attempt to prove, from nitrated filver,
(lapis infernalis), corrofive fublimate,
and other metallic falts, that the natural acrimony of acids is heightened
by combination with certain bafes, the
opinion, if we examine into the matter
clofely, will appear to be groundlefs.
It is clear, from an hundred inftances,
that the acrimony of acids is diminifhed in proportion to their faturation;
to fuppofe it increafed is repugnant to
the

the nature of the thing : befides, an extraordinary degree of vehemence is afcribed to thefe corrofive falts, becaufe they attack animal bodies when they come in contact with them. But the true caufe of the corrofion is the dephlogiftication of the metallic bafes, (XV. XVII.), which are for this reafon capable of tearing away the inflammable principle contained in animal fubftances, for to this they always tend with great force ; and thus they in fome meafure moderate their ftrong attraction, for the acids are infufficient to faturate them. The various hypothefes concerning the origin of the nitrous and marine acids from the vitriolic, are well known, but they remain to this day unfupported by any valid argument, as will alfo hereafter appear, if I miftake not, with refpect to the fluor acid.

THE

THE preceding acids attract alkalis in preference to earths ; but here a different order begins. Fluor acid, faturated with vegetable alkali, is decompofed by lime-water, and yields fluor and uncombined alkali. The acids which abound in phlogifton feem, for the moft part, to prefer lime to alkalis. Ponderous earth, faturated with fluor acid, is foluble in a large quantity of hot water, and, upon the addition of lime-water, yields its folvent to the lime, an effect eafily afcertained, for the liquor is rendered turbid, and fluorated lime is depofited. Fluor acid feems to take magnefia from vitriolic acid, but this does not hold with refpect to lime ; hence, therefore, the firft place might be affigned to magnefia ; but when I afterwards repeated the experiment, with all poffible care, there was no appearance of precipitation. The fediment, therefore, in the firft experiment, was perhaps filiceous earth, which

which quitted the acid when it came to be diluted in the folution.

In the dry way, the order of this acid is made the fame with the preceding, though it remains to be determined by experiment. It is certain, however, that fluor mineral is not decompofed by cauftic fixed alkali, (Scheme 51.), though aerated alkali effects a decompofition; but it is in confequence of a double elective attraction, (Scheme 63.).

XX.

Column Ninth, the Arfenical Acid.

That admirable difcovery, which difclofed the compofition of marine acid, at the fame time points out a method of acquiring pure acid of arfenic. Macquer's arfenical falts indeed fhew evidently the acid nature of white arfenic;

fenic : the pure acid, however, could not be feparated before the method of dephlogifticating the muriatic acid was brought to light. We are now acquainted with two ways. In the firft, one part of pulverized black magnefia, and three parts of acid of falt, of which the fpecific gravity to that of water, fhould be as 5 : 4, are mixed in a tubulated glafs retort, having a bulb capable of containing four times the quantity of the ingredients : to the retort a receiver is adapted, containing one-fourth of pulverized white arfenic, together with one-eighth of diftilled water : the retort is to be heated in a fand bath, and the manganefe will quickly dephlogifticate the marine acid, which again regains its complement from the white arfenic. The acid of falt being thus regenerated, unites with a portion of the water, and diffolves part of the arfenic : the reft of the water is feized by the arfenical acid, as it

is

is gradually dephlogifticated; fo that the liquor of the receiver is divided into two ftrata, and in a few hours all the arfenic difappears : at this period the liquors fhould be diftilled to drynefs in a retort. That which is collected in the receiver, confifts of butter of arfenic, and marine acid unmixed ; but the white refiduum in the retort, which fhould be heated red hot to free it completely from acid of falt, exhibits real acid of arfenic in a folid form, (Scheme 17.), which is eafily foluble in water.

THE other method, is as follows : let two parts of pulverized white arfenic be diffolved in a tubulated glafs retort, in feven parts of marine acid, by flow boiling. Let the liquor collected in a receiver luted to the retort, be poured back, and at the fame time three and one-half parts of nitrous acid, of the fame fpecific gravity as the abovementioned marine, be added ; then let a

receiver

receiver be adapted again, but without a lute. By the affiftance of the heat, the nitrous acid feizes the phlogifton of the arfenic, and emits red fumes; but the diftillation muft be carried on till no more of thefe fumes are feen. Then, one part of white arfenic is to be added, which fhould be in like manner diffolved by gentle boiling, and afterwards one and one-half part of nitrous acid, which dephlogifticates the diffolved arfenic with effervefcence, and red fumes arife. Laftly, diftil to drynefs, and the refiduum, after flight ignition, will be found to confift of pure arfenical acid, which is fixed in the fire, attracts moifture in the open air, and is foluble, if it be fufficiently dephlogifticated, in twice its weight of water. It fhould be freed by thorough wafhing in a filter from the filiceous powder, which comes from the corrofion of the glafs during ignition.

ARSENICATED

ARSENICATED vegetable alkali is immediately decompofed by lime-water, and the alkali is difengaged. I have fcarce any doubt but ponderous earth and magnefia prevail over alkaline falts; though I muft confefs that this conjecture has not been yet confirmed by experiment.

IF acid of arfenic does not diffolve metals in their complete ftate, it at leaft diffolves them when calcined to a due degree. It is, moreover, to be obferved, that no inflammable air is generated during the folution of iron; for the phlogifton, being abforbed by the acid itfelf, regenerates white arfenic.

XXI.

XXI.

Column Tenth, Acid of Borax.

THE fubftance commonly called fedative falt, is more nearly allied to a-
cids than any other clafs of bodies.   It
reddens turnfole ; faturates alkalis and
foluble earths.   It alfo diffolves various
metals, and has other properties which
fhew its acid nature ; and it feems better entitled to the name of acid of borax, than that of fedative falt.

DEPURATED borax may be decompofed by boiling with lime ; the acid forfakes the cauftic foffil alkali to feize
the lime, and produces a falt fcarce foluble.   That the fame thing takes place
with vegetable alkali, faturated with
acid of borax, is hitherto only a pro
bable conjecture ; as alfo on the addition

tion of ponderous earth and magnefia.

Acid of borax attacks metals with
difficulty. The eafieft way to combine
thefe fubftances is by a double affinity ;
but, to avoid miftakes, the borax fhould
be faturated with fedative falt, of which
there is required fomewhat above an
equal weight, before the reaction of
the alkali entirely ceafes. I have
dropped a folution of fuch borax into
metallic folutions, freed as much as
poffible from fuperfluous acid. Gold,
platina, bifmuth, and manganefe, diffolved in their proper menftrua, remained undifturbed ; but folutions of
mercury, lead, copper, iron, tin, nickle,
cobalt, and zinc, were immediately
rendered turbid, and yielded metallic
falts of very difficult folubility, (Scheme
28.).

XXII.

XXII.

Column Eleventh, Acid of Sugar.

Most vegetables fhew manifeft figns of acidity in their fruit or juices ; and as the very few vegetable acids, which are known with tolerable accuracy, have different properties, the diligence of pofterity will certainly bring to light a great number. The chief obftacle which prevents us from becoming acquainted with them, is the great difficulty of purifying them; for they are fo involved in other fubftances, as fcarce to admit of being extricated. I produce but a few here, of which the greater part labour under the imperfection of being deftructible by fire. Do all of them agree in their primary principles? Can they be tranfmuted? Thefe queftions muft be determined by

accurate

accurate experiments, made with this
view. Of the acids of tartar, lemon,
and milk, it is certain, that they all,
upon the addition of fpirit of wine and
water, and after a digeftion of feveral
weeks, are changed into vinegar. At prefent they muft be confidered as different,
fince they can always be obtained perfectly the fame, and poffefs properties
perfectly diftinct, conftant, and of the
utmoft importance in chemiftry.

THE acid which exifts in fugar, is
found to be fo clofely united with an
oily matter, that it has never yet been
poffible to feparate it, but by the nitrous acid which deftroys that matter.
For this purpofe, let fix—eight parts of
ftrong nitrous acid be poured upon one
of white fugar, reduced to powder, in
a glafs retort, and be gently boiled.
The nitrous acid in a fhort time feizes
the phlogifton, and emits red fumes;
after the ceffation of which, the liquor
remaining

remaining in the retort ſhould be poured into a large veſſel, and it will afford
cryſtals of a priſmatic form, and very
acid taſte. If the lixivium be dephlogiſticated by two—four parts of nitrous
acid, it will again depoſit cryſtals of
inferior purity indeed, but they may
be purified by ſolution and repeated
cryſtallization. This acid may be alſo obtained from honey, gum arabic,
and ſpirit of wine, by means of the nitrous acid, but in ſmaller quantity.
It poſſeſſes all the properties of acids
in general ; and beſides theſe, ſeveral
peculiar to itſelf, by which it is di-
ſtinguiſhed from all others. It totally
differs from nitrous acid, and in many
reſpects is of an oppoſite nature ; ſo
that its origin cannot with any degree
of probability be aſcribed to that acid.
But this queſtion is conſidered at greater length elſewhere * ; the attractions

are

* Opuſc. vol. i. p. 251.

are the chief object of the prefent dilfertation.

IT attracts lime moft ftrongly, and forms with it a faline combination, infoluble in water, whence we may eafily perceive the neceffity of lime-water in the refining of fugar. The juice of the fugar-cane has an excefs of acid, which prevents the concretion of the fugar. For if this acid be added to a folution of perfect fugar, it will not yield cryftalline grains, but a glutinous mafs. Nothing can therefore be of greater fervice than lime-water, which not only abforbs the uncombined acid, but likewife forms an infoluble falt, that either falls to the bottom or floats in the froth. Alkalis indeed faturate this acid, but they form falts which can fcarce be feparated on account of their folubility.

PONDEROUS

PONDEROUS earth, magnefia, and alkalis, yield this acid to lime. It attacks almoft all the metals. Its com-
parative power, with refpect to other
acids, will foon be feen in the columns
of alkalis, earths, and metals. This
acid is an excellent teft for detecting
lime any way diffolved or fufpended
in water, for the fmalleft drop of a folution of it immediately feizes the lime,
forming with it a white infoluble powder, which falls to the bottom. If we
know the quantity of lime in a given
weight of this faline powder, we eafily
learn, at the fame time, the quantity of
lime, in any cafe, from the fediment,
completely precipitated from a determinate quantity by means of this acid,
carefully collected, wafhed, and weighed.

IT cannot fuftain experiments in the
dry way. The cryftals alone, being
expofed to fire in a glafs retort, partly
indeed

indeed fublime, almoft unaltered, but
the greater part is refolved into an acid
liquor, which will not cryftallize, as
alfo happens to the fublimate, if it be
again fubjected to the operation. During this deftruction of the acid, a great
quantity of inflammable air and aerial
acid is extricated. The acid of fugar
undoubtedly abounds with unctuous
matter, yet it is of a very fubtile nature, for in the fire it leaves no traces
of charcoal or foot.

XXIII.

Column Twelfth, Acid of Tartar.

I HAVE above explained the nature
of tartar, (IX.), and fhall now briefly
mention the procefs for obtaining the
acid pure. To an hundred parts of
cream of tartar, diffolved in boiling water in a tin boiler, let fmall quantities
of

of chalk, wafhed, dried, and pounded, be added at intervals. This muft be continued as long as any effervefcence is excited by the addition ; about twenty-eight parts will be required. When the point of faturation has thus been attained, let the liquor be decanted and evaporated to drynefs ; it will yield fifty parts of tartarized vegetable alkali. The powder remaining in the bottom is lime faturated with the abundant acid of tartar, which, when wafhed and dried, amounts to above triple the chalk ufed, *viz.* an hundred and three. Let this compound be put into a phial, and let there be gradually added three hundred parts of vitriolic acid, containing two hundred and feventy of water, and thirty of the ftrongeft acid. Let the mixture be digefted for twelve hours, and often ftirred with a wooden fpatula. Laftly, let the clear liquor be poured off, and the refiduum wafhed till it has loft its acid tafte ; let the wa-

ter

ter be filtered and added to the decanted liquor. We have here a folution of
acid of tartar, which, if evaporated to
drynefs, affords near thirty-four parts of
a cryftalline mafs. To try whether it
contains any vitriolic acid, let one or
two drops of a folution of fugar of lead
be dropped into the diluted folution, a
white fediment wiil immediately fall
down, upon which concentrated vinegar fhould be poured, and it will foon
difappear, if it confift of lead, faturated
with acid of tartar; but vitriol of lead
will not be diffolved. Should fome of
this be detected, it fhews that the acid
of tartar is adulterated; but it may be
eafily purified by digeftion with a fmall
quantity of tartarized lime. On the
other hand, if too little vitriolic acid
has been added, fome acid of tartar will
remain in the refiduum, which is eafily
tried by throwing it on red hot coals;
for pure gypfum neither grows black
in the fire, nor emits an odour of fpirit

of

of tartar, which, however, happens, if
it be contaminated with tartarized
lime. A filtered folution of acid of
tartar evaporated to drynefs will gradually depofit cryftals in a cool place,
which generally confift of divaricating
lamellæ.

IF newly burned lime be ufed in the
place of chalk, twice the weight of tartar will be totally decompofed, and
therefore there will be uncombined vegetable alkali in the liquor ; but aerated lime can only abforb the excefs of
acid, and the tartarized tartar will remain unaltered. An hundred pounds of
purified tartar contains about twenty-
three pounds of pure vegetable alkali,
forty-three of faturating, and thirty-
four of abundant acid. The difference
of wines, and of proceffes for purification, will doubtlefs occafion fome difference,

rence, but I have not yet been able to afcertain the limits.

FROM what has been faid, it appears that lime, with refpect to acid of tartar, has the fuperiority over vegetable alkali, which is alfo true of ponderous earth and magnefia ; for thefe earths, when deprived of aerial acid, and added to an exactly neutral folution of tartarized vegetable alkali, in a few minutes attract the acid of tartar in a heat of digeftion ; and evident figns of uncombined alkali appear on adding the proper tefts. Mineral and volatile alkali exhibit the fame phænomena. To afcertain whether lime or ponderous earth prevails, I dropped lime-water into a folution of tartarized ponderous earth, by which it was quickly rendered turbid, fo that every doubt with refpect to the fuperior attraction of lime is removed.

CRYSTALLIZED

CRYSTALLIZED acid of tartar, when expofed to fire, immediately grows black, and yields a fpongy charcoal, which, however, foon turns white, if heated to incandefcence, and is much contracted in bulk. When diftilled from a glafs retort, it affords, in the receiver, a phlegm fcarce acid, together with oil; the carbonaceous refiduum confifts of a portion of earth, which fhews no figns either of alkali or acid. Hence, therefore, it appears, that there exifts oil in this acid, which is deftroyed in the fire, and is by no means convertible into fixed alkali. Treated with nitrous acid in the method above defcribed, (XXII.), it afforded no acid of fugar, and could not indeed be fo much deprived of oily matter, as not to grow black in the fire. Some perfon is faid, however, to have fucceeded in converting acid of tartar into acid of fugar, which I have not yet been able to confirm by a repetition of the experiment.

riment. Acid of tamarinds, and perhaps of berberries, feem to agree in
all refpects with that under confideration.

XXIV.

Column Thirteenth, Acid of Sorrel.

SALT of forrel is vegetable alkali faturated in excefs with a peculiar acid,
and therefore of the fame nature as tartar. To obtain this acid pure, is a
work of difficulty. The vitriolic, nitrous, and muriatic acids, attract indeed
the bafis, but it is very hard to free the
acid, thus fet loofe, from all impurities.
An hundred and thirty-feven parts of
chalk perfectly decompofe an hundred
of this falt, attracting both the faturating and exceffive acid. The clear liquor yields thirty-two parts of aerated
vegetable alkali; nearly the fame quantity

tity as is obtained by fire. The fediment contains *forrelled* or *oxalited*
lime, which, after edulcoration and ex-
ficcation, weighs an hundred and feven-
ty-three. The component parts of this
falt cannot be feparated like thofe of
tartar, for the acid of forrel takes lime
from the vitriolic. Mr Scheele difcovered another procefs. He added the
abundant acid of falt of forrel, faturated with volatile alkali, to a nitrous fo-
lutibn of terra ponderofa ; an exchange
of principles immediately took place,
in confequence of a double elective attraction ; and the ponderous earth, with
the acid of forrel, a compound difficult of folution, fell to the bottom,
(Scheme 25.). This fediment is decompofed by acid of vitriol, which prefers ponderous earth to every fubftance
yet known ; and the acid defired may
be poured off; but as it has been but
lately difengaged, its properties have
been little examined. Meanwhile, it
feems

feems more nearly allied to acid of fugar than of tartar ; it differs, however,
from both, for the abundant acid of forrel forms, with vegetable alkali, falt of
forrel, analogous to tartar, but decrepi-
tating in the fire, fufible, fcarce turning
black, and capable of being totally decompofed by aerated lime,—properties
not belonging to tartar.   Now, faccharated vegetable alkali agrees neither
with tartar nor falt of forrel.

I know with certainty that acid of
forrel prefers lime and ponderous earth
to alkaline falts ; but it is as yet doubtful whether this be true of magnefia.

This acid is alfo deftructible by fire,
but it neither fwells fo much, nor turns
fo black as acid of tartar ; by diftillation, a phlegm, far more acid than that
which is feparated from tartar by this
procefs, but no oily matter, is obtained.
Genuine falt of forrel, as it is found in
the fhops, varies not a little in the proportion

portion of its proximate principles. An hundred parts of fome fpecimens yield, on combuftion and elixation, but thirty-one of vegetable alkali, while, to faturate the excefs of acid, there is required an hundred and twelve and one-half. Others afford above thirty-feven, and require not above eighty-feven for faturation. It is obvious, that in the former cafe the faturated acid is to the abundant as $1 : 3\frac{1}{2}$, but in the latter as $1 : 2\frac{1}{2}$. The abundant portion is always larger than the faturated, which explains the ftronger acidity of this falt compared with tartar. But the difference perhaps depends folely on the place where the forrel grows, or on the method by which the falt is extracted. The acid of forrel, (or more properly, perhaps, the *oxaline* acid), when procured by diftillation, precipitates filver, quickfilver, and lead, from the nitrous acid ; with alkalis, forms cryftallizable compounds, and, by evaporation without

out addition, may be reduced to acid
cryftals.

XXV.

Column Fourteenth, Acid of Lemon.

ALTHOUGH the acid of lemon has
been long known, it has been but fuper-
ficially examined. The expreffed juice
is to be freed from mucilage and fœcu-
lent matter, by a long fubfidence, and
fhould then be expofed to cold, that the
watery part may be frozen. It feems
to contain a little vegetable alkali, for
upon dropping in acid of tartar, in a
few days a fmall quantity of tartar is
found at the bottom. Stahl affirms
that acid of lemon, faturated with
crabs-eyes, and with the addition of
a little fpirit of wine, infenfibly affumes, if it be kept in a phial flightly
ftopped, the nature of vinegar. An
equal

equal quantity of the fame juice of lemon requires, for its faturation, of pure vegetable alkali fixty-nine parts, of pure mineral fifty-one, and of volatile twenty-five. Thefe compounds do not eafily cryftallize.

IT remains to be determined, by a greater number of experiments, whether acid of lemon certainly prefers lime, ponderous earth, and magnefia, to alkalis.

THE acids of fluor, arfenic, borax, fugar, tartar, forrel, and phofphorus, agree in this refpect, that if they be faturated with earths they are fcarce foluble in water; and, therefore, if fmall quantities of earth be added, the folutions remain clear, on account of the excefs of acid, but near the term of faturation they become a little turbid; and when this is attained, all that was diffolved is inftantly precipitated, and can fcarce

be

be again diffolved but by excefs of
acid. The fame thing is perhaps true
of fome other acids, but not of all.

XXVI.

Column Fifteenth, Acid of Benzoin.

THE refin, which, in the fhops, paffes
by the name of *benzoin*, or *affa dulcis*,
contains an acid falt, which may be
feparated by fublimation, and concretes
into cryftalline filaments or fpiculæ.
It may be acquired in the humid way
more free from oily matter, by extracting the acid with lime, and then
feparating the compound with muriatic
acid *. The acid of benzoin being diflodged, falls to the bottom, uncombined, and more difficult of folution
than when united with the lime.

BOILING

* Scheele in the Stockholm Tranfactions, 1775.

BOILING water takes up one twenty-fourth of its weight; but at the ordinary temperature fcarce more than one five-hundredth. The tafte, approaching to that of fugar, gives a flight fenfation of acrimony, but none of acidity. It, however, reddens tincture of turnfole.

SPIRIT of wine readily diffolves it, even without heat.

IT is totally refolved by fire into white vapours: it is, however, fufible with fmoke, on the fudden application of a fufficient heat; but can fcarce be inflamed without the actual contact of fome burning fubftance.

IT readily combines with alkaline falts; but none of the compounds, except that with mineral alkali, afford cryftals, that are not liable to deli-quefcence.

LIME-

Lime-water, dropped into the folutions, precipitates the alkaline bafes, as alfo magnefia and clay. Lime, therefore, exceeds them in attractive power, but it is unable to feparate ponderous earth; nor does the latter, when burned and diffolved in water, precipitate lime united with acid of benzoin; fo that here a combination of three ingredients feems to take place.

The earthy falts, containing acid of benzoin, are with difficulty diffolved in water, efpecially thofe that contain the ponderous and calcareous earths.

XXVII.

Column Sixteenth, Acid of Amber.

A crystallized volatile acid may be obtained from amber, by diftillation, together with an acetous liqour and an oil.

oil. The acid may be in some measure purified by solution, and a second
cryftallization. That which I used in
the following experiment was thus prepared; it shews evident signs of acidity.

COMBINATIONS of this acid with
alkalis may indeed be made to cryftallize, but they are all deliquefcent, except
that into which the mineral alkali enters. Lime and ponderous earth afford falts of difficult folubility; clay
yields cryftals, and magnefia a compound like gum.

THE metals, when duly dephlogifticated, are foluble in this acid, and for
the moft part they afford permanent
cryftals.

PONDEROUS and calcareous earths,
and magnefia, take this acid from alkalis. Ponderous earth alfo precipitates

tates fuccinated lime and magnefia ;
lime-water makes no change in the folution of fuccinated ponderous earth,
but it diftinctly prevails over magnefia.

XXVIII.

Column Seventeenth, Acid of Sugar of Milk.

BARTHOLET firft mentioned fugar of
milk in the year 1619 *, but no one
till Mr Scheele gave a complete analyfis of it †.   An hundred parts of this
falt yield fifteen and one half parts
of acid of fugar, and of another acid
hitherto found only in fugar of milk,
and which is here to be confidered
about twenty three and one half.

THIS

---

* Encyclop. Hermetico-dogmatica.
† Stockh. Tranf. 1780.

THIS acid is obtained in the form of a white powder, and is not eafily foluble in water, for fixty parts of boiling water take up but one of the acid. The folution turns turnfole red.

IN combination with alkaline falts it forms compounds far more foluble indeed, but ftill requiring a weight of boiling water many times exceeding the folvend. The compounds formed by this acid and earths, are fcàrce foluble at all. As to attractions, alkalis obferve the ufual order. Ponderous and calcareous earths, and magnefia, are fuperior to alkalis, but it is difficult to determine the fuperiority among them, as they are fcarce foluble.

XXIX.

XXIX.

Column Eighteenth, Diſtilled Vinegar.

THIS acid, produced, during fermentation, from the preceding, which may
in ſome meaſure be denominated crude,
differs from them in the great infe-
riority of its attractive power for
earths and metals ; but, on the other
hand, exceeds them in ſubtilty, not
being altered by diſtillation. Vinegar
alſo contains phlogiſton, but more intimately combined, or at leaſt more
concealed. It has been mentioned a-
bove that acid of tartar, digeſted with
water and ſpirit of wine, affords vinegar. B. de Vigenere long ago obſerved, that cryſtals of tartar ſometimes
form in vinegar *. An acid alſo, very
nearly reſembling vinegar, may be procured by diſtillation from guaiacum
and

* Du feu & du ſel, cap. 35. 1608.

and birch, from wax, fugar and amber, which is not to be confounded with acid of fugar, and falt of amber.

PURE lime cannot detach vinegar from fixed alkali, and it therefore feems to refemble the vitriolic, nitrous and muriatic acids in its attractions; and I have difpofed them in the fame order, though the priority of ponderous earth has not yet been confirmed by experiments.

I CONJECTURE, that, in the dry way, the fame feries, as in the fifth column, prevails, till I have leifure to confirm or correct this opinion by experiment.

XXX.

XXX.

Column Nineteenth, Acid of Milk.

To obtain the acid of milk, curdled by fpontaneous acefcency, let the whey be collected and evaporated to one-eighth. This refiduum is to be faturated with lime, that the phofphorated lime may be feparated; then let the calcareous earth that has been diffolved be precipitated by acid of fugar; and at laft by highly rectified fpirit of wine, the acid of milk is obtained free from the phofphorated lime, fugar of milk, vegetable alkali, and mucilage which milk contains *.

THIS acid feems to be intermediate between vinegar and acid of ants; it, however, exceeds vinegar in attractive power. By the addition of fpirit of
wine,

* Scheele, Act. Stockh. 1782.

wine, it is really changed into vinegar, after a month's digeftion.

WITH the alkalis, it forms delique-fcent falts, and alfo with the earths, a-mong which magnefia, contrary to what would be expected, forms the moft permanent combination. Scarce any metal, befides zinc, forms cryftals with this acid ; not even lead, which yields a fweet folution, and depofits fome vi-triolated lead.

WITH refpect to alkalis and earths, the order of attractions is the fame as that of vinegar.

XXXI.

Column Twentieth, Acid of Ants.

THIS acid approaches very near vi-negar ; they differ, however, in many refpects.

respects. The former forms with magnesia, zinc, and iron, cryftallizable
falts ; the latter only deliquefcent. Its
combination with magnesia is peculiarly remarkable *.

ITS attractions have hitherto been
very imperfectly examined ; in the
mean time, as far as has hitherto appeared, this acid obferves the fame order as
vinegar ; which, however, is weaker, as
the following columns fhew.

ALL the acids of vegetables, as well
as that of ants, may be totally refolved
into an elaftic fluid, confifting partly
of aerial acid, and partly of inflammable air.

XXXII.

* Opufc. vol. ii. p. 389.

XXXII.

Column Twenty-firſt, Acid of Fat.

The celebrated Crell, by repeated diſtillations of fat, obtained from 2 lib, $14\frac{1}{8}$ ounces of oil, of charcoal $10\frac{19}{24}$, and of acid $7\frac{1}{4}$ *.

The ſalts formed by this acid, ſaturated with alkalis and earths, are very like thoſe that vinegar generates with the ſame baſes. Alkalis yield it to earths. A compariſon of this with other acids will be found in the ſequel.

XXXIII.

XXXIII.

Column Twenty-second, Phosphoric Acid.

ANOTHER acid which pervades all the
kingdoms of nature, is especially obtained from human urine, by collecting the substance, usually called microcosmic salt, and fusing it in the fire : but
it is to be observed, that this is a triple
salt, and always contains two acids, the
phosphoric and *perlatum,* which shall be
immediately more fully described, together with volatile alkali.   When the a-
cid is obtained in water by slowly burning phosphorus in the air, it abounds
with phlogiston ; which however, in an
open vessel, is gradually dissipated : it
thus becomes much fitter for ascertain-
ing attractions than the former ; for the
presence of perlate acid in that, may
give rise to a diversity in the decompositions.

THE

THE acid of phofphorus prefers lime
to alkalis ; and therefore alkalis united
with it, are immediately rendered turbid by lime-water, and a faline powder,
very difficultly foluble in water, is gradually depofited, confifting of lime faturated with phofphoric acid. As phofphorated vegetable alkali can receive an
excefs of acid, I fufpected that this excefs
only might produce the precipitation with
lime-water, and that the neutral compound might remain entire ; but I decompofed the whole with a fufficient
quantity of lime-water, and nothing
but an alkaline liquor was left. Phofphorated lime is indeed precipitated
by alkali, but not in confequence of a
fuperior elective attraction. This falt
is not foluble in water without an
excefs of acid ; and whatever takes a-
way the medium of folution, as in the
prefent cafe the alkali does by faturating it, caufes a precipitation of phofphorated lime : aerated alkali, however,

ever, decompofes it by means of a double attraction ; and aerated lime falls down, (IX.). In ambiguous cafes of this kind, we muft form our conclu-fion from the decompofition of that falt which is more eafily foluble. Thus, the prefent queftion is anfwered by the precipitation of phofphorated alkali by lime-water, and not by that of phofpho-rated lime by alkali. The fuperior power of magnefia and terra ponderofa is not yet afcertained with fo much certainty.

In the dry way, I place lime before magnefia and ponderous earth, fince the former undoubtedly difpoffeffes fix-ed alkalis, which is not yet certain concerning the latter.

XXXIV.

XXXIV.

Column Twenty-third, Acidum Perlatum.

In 1740, Haupt found a falt in human urine, and defcribed it under the
title of *fal mirabile perlatum*; it is the
fame which Margraaf mentions as altogether unfit for generating phofphorus
with powdered charcoal. Neverthelefs, many have fuppofed, that this falt,
which occurs in urine with the microcofmic falt, contains the phofphoric a-
cid, and have explained its rejection of
phlogifton from the fuperior attraction
of foffil alkali. But Mr Prouft has
lately difengaged the fubftance, which a-
lone performs the office of an acid in *fal
perlatum*; in microcofmic falt, indeed, it
is united with the phofphoric acid. He
digefted fal perlatum in diftilled vinegar, brought it to afford cryftals, and
precipitated from the mother ley, by
fpirit

spirit of wine, a thick liquor, which, after being well washed in spirit of wine, and then diffolved in diftilled water, yielded an acid * which I call the *perlate*, till a more accurate examination shall fuggeft a better name. Mineral alkali, if I miftake not, covers the acid. The cryftallization, tafte, reaction, and efflorefcence, point out an excefs of alkali. Perhaps this excefs prevents spirit of wine, which rejects no acid, from extracting the acid part. It fufes with ebullition, and appears pellucid, but when cooled it becomes opaque: it is taken up by acids, and may be feparated by spirit of wine : it exifts with the phofphoric acid, as well in bones as in microcofmic falt.

It certainly prefers lime, ponderous earth, and magnefia, to alkalis. With mineral alkali, it regenerates fal perlatum.

It

* Rozier's Journal.

IT would be a peculiar and highly remarkable phænomenon, if a neutral compound, by means of an intimate union with phlogiſton, could act as an acid without decompoſition. This, however, ſeems to be the caſe with ſedative ſalt, and ſtill more certainly with the acid of Pruſſian blue, which will be deſcribed below. That obtainable from ſal perlatum, is, if I miſtake not, of the ſame nature. By means of a ſtrong impregnation of phlogiſton, it is ſo cloſely combined with a certain portion of foſſil alkali, that we have not as yet been able to effect a ſeparation of the principles ; and what well deſerves notice, the compound, as a ſimple acid, takes up alkalis, earths, and metals, though ſeveral properties plainly indicate an exceſs of alkali. Meanwhile till their nature is better aſcertained, I conſider them as acids, ſince they ſeem to approach neareſt to theſe ſubſtances.

XXXV.

XXXV.

Column Twenty-fourth, Acid of Prussian Blue.

I have long conjectured, not without reason, that the tinging matter in Prussian blue is of an acid nature, as it forms compounds of an intermediate kind with alkaline salts, as well as with earths and metals. Mr Scheele has lately taught us how to separate the acid in a pure state *. Phlogisticated alkali, as it is commonly called, is a triple salt; containing the tinging acid, saturated partly with iron, and partly with alkali. This salt, boiled in a retort with weak vitriolic acid, emits the tinging acid in an inflammable aerial form, which may be absorbed by water placed in the receiver. And as at the same time, some vitriolic acid passes

* Act. Stockh. 1782.

paſſes into the receiver, the liquor ſhould be again diſtilled with a little chalk, till one-fourth ſhall have paſſed over; which is a ſolution of the preſent acid in water. The following procefs anſwers the ſame end with leſs trouble: let ſixteen parts of Pruſſian blue be boiled in a cucurbit, with eight of mercury, calcined by means of nitrous acid, and forty-eight of water, for a few minutes, with conſtant agitation. The mixture becomes of a cineritious yellow; it ſhould be put on the filter, and the reſiduum elixated with boiling water. To the filtered liquor, let twelve parts of pure iron filings be added, and three of concentrated vitriolic acid. After a ſhaking of ſome minutes, the whole maſs is turned black by the reduced mercury. After the ſubſidence of the powder, the clear liquor is to be decanted into a retort, and one-fourth abſtracted.

THIS

THIS acid, in fome refpects analogous to the perlate, is diftinguifhed by
a peculiarly difagreeable tafte and fmell,
and confifts of aerial acid, volatile alkali, and phlogifton. It fpeedily flies off
in an open veffel : it feems to prefer alkalis to earths. Its forcible attraction
for metals will be more particularly confidered in the fequel.

XXXVI.

Column Twenty-fifth, the Aerial Acid.

OF this acid, which is common to all
the kingdoms of nature, I have treated
at length in another effay, and have
particularly examined its attractions in
§ 20. *. I fhall now add only a fhort
comparifon of it with phlogifticated vitriolic acid, fince they are wretchedly
confounded by fome, who find it more
eafy

* Opufc. vol. i. p. 43.

eafy to burden natural philofophy with feigned hypothefes, than to enrich it with accurate experiments. I have faid above, that phlogifton, united with vitriolic acid, in different proportions, produces either vitriolic acid air, which, when abforbed by water, is called phlogifticated vitriolic acid, or fulphur, (XIII.).

PHLOGISTICATED VITRIOLIC ACID has a moft penetrating fmell ;—faturated with vegetable alkali, generates the fulphureous falt of Stahl ; the cryftals fpiculæ indiftinctly hexangular ; — they detonate with nitre ;—may be totally fublimed by a proper application of heat ;—do not effervefce with other acids ; — gradually lofe their phlogifton, and are at laft fpontaneoufly changed into vitriolated tartar.

THE AERIAL ACID has no fmell ;—with vegetable alkali, forms aerated alkali ; the cryftals quadrangular prifms ;—they do not yield a fpark with nitre ;—are fixed ;—emit conftant numberlefs air-bubbles, till the faturation is complete, even with phlogifticated vitriolic acid, to which the aerial acid is inferior in attractive power ; — remain unchanged, unlefs fire or fome ftronger acid expel the weak aerial acid.

HENCE

HENCE let thofe who are fkilled in chemiftry, and regard truth, form a judgment. If any one fuppofes other combinations of vitriolic acid with phlogifton, befides thofe mentioned above, he muft prove the mode of their formation, not by opinions, but by experiments.

A SIMPLE experiment proves that the aerial acid is attracted by atmofpheric air; for if a phial filled with the former be fet in an open place, where the ambient fluid undergoes no agitation, it will be found to contain atmofpheric air only. Nay, aerated water yields its volatile acid to the atmofphere.

XXXVII.

XXXVII.

Column Twenty-fixth, Cauftic Vegetable Alkali.

HAVING examined the attractions of the acids, we now come to the alkalis, which are commonly combined with aerial acid, but when freed from this, are called cauftic, or even pure ; and it is in this ftate only that they can be employed for the prefent purpofe, for, when aerated, they give rife to double affinities ; fee Schemes 1—8. 32—36. 46. 51. 62. and 63. Thofe which are properly called faline, are of three kinds. The firft is denominated vegetable alkali, being all obtained from this kingdom ; and of this only I fhall fpeak in the explanation of column twenty fixth.

THE

THE vegetable alkali adheres moſt
ſtrongly to vitriolic acid, for it not only is not diſlodged by either of the
others, but takes this acid from them.
In what manner nitre and ſea ſalt effect a partial decompoſition of vitriolated tartar, and the acid of tartar, likewiſe of nitre and digeſtive ſalt, I have
explained above, (IX.), and in Scheme
9. 10. and 11. have ſymbolically re-
preſented the partition of the acid by
darts.

THE ſecond place belongs to the nitrous, (VI.), and the third to the marine acid. Whether the aeriform muriatic acid is capable of decompoſing
nitre, has not been ſufficiently tried.
In order to account for the order of
the other acids, it will be proper to
mention the chief experiments.

THE celebrated Crell has admirably
ſhewn, that the acid of fat is neither
expelled

expelled by the acids of fluor nor of phofphorus. That of fluor dropped into a folution of phofphorated vegetable alkali, immediately precipitates a triple falt, confifting of fluor acid, vegetable alkali, and flint *. I dropped arfenical acid into a folution of phofphorated vegetable alkali; after the mixture had ftood twenty-four hours, fpirit of wine precipitated the phofphoric falt in no refpect altered, and the acid of arfenic remained in the fpirit. Therefore the phofphoric is the ftronger. The acid of tartar decompofes all the falts which contain vegetable alkali, as far as I have tried, as the combinations of this bafis with the vitriolic, nitrous, marine, faccharine, phofphoric, and arfenical acids, fome completely, but others only in part, (IX.). To difcover, therefore, its real power, acid of tartar fhould be dropped into folutions of mineral alkali, combined with the feveral acids.

---

* Opufc. vol. ii. p. 34. 37.

acids. For this alkali attracts the acids
in the fame order as the vegetable, without acquiring any excefs, by which the
obferver might eafily be mifled. The
acid of fugar feems more powerful than
that of tartar ; thofe of forrel, lemon,
and amber, weaker ; but their order
with refpect to each other has not yet
been fufficiently afcertained. Next follow the acids of ants, milk, and benzoin, which are ftronger than vinegar.
Acid of borax is expelled by vinegar,
as alfo the perlate acid, the nitrous and
vitriolic fully phlogifticated. The aerial yields almoft to all ; it however
precipitates folutions of flint, fulphur,
and oil ; it even expels the acid of
Pruffian blue. Vegetable alkali takes
up copper and tin, but their places are
uncertain.

In the dry way, the acids of phofphorus, borax, arfenic, and perlatum, are
fuperior on account of their fixity,

(IV.) :

(IV.): the reft, except thofe that are deftroyed by ftrong heat, obferve the fame order as in the moift way. Pure earths coalefce with alkalis in heat, but in what order is uncertain; nor can it be eafily afcertained, fince when any one is united by fufion with alkali, it is not precipitated by the addition of another, but both combine with the menftruum, and form an homogeneous mafs. The fame thing holds with re-fpect to fulphur.

XXXVIII.

Column Twenty-feventh, Cauftic Mineral Alkali.

THIS fixed alkali has received the name of mineral, from the great quan-tity of it which occurs in the mineral kingdom; it is alfo found in plenty in fome vegetables growing in the fea,

or

or in a falt foil. It differs from the preceding, being both weaker, (XII. XIV. XVI.), and generating different falts with the fame bafes. The acid of tartar does not in any refpect change fea falt, (Scheme 12.), which affords an eafy method of diftinguifhing it from digeftive falt. For the acid of tartar is the beft teft yet known for detecting the prefence of vegetable alkali in any mixture, as it decompofes all the falts which have this bafis, fome totally, others only in part, (IX.). This decompofition is denoted by the depofition of tartar, which, if the dofe of alkaline falt be confiderable, and the folution concentrated, becomes perceptible in an inftant: a fmaller quantity, efpecially in a weaker folution, requires the fpace of a day or two.

As to the order of attractions reprefented in this column, I have not found it to differ from that in the thirtyfeventh.

feventh. This acid more certainly
fhews the true order with refpect to
acid of tartar, and in certain cafes
more diftinctly, as the compounds are
more difficult of folution than thofe
containing the vegetable alkali. This
is often advantageous in determining
the attractions. Scheme 13. fhews
the decompofition of borax by nitrous
acid; Scheme 42. of fea falt by the
fame, and 43. by acid of arfenic.

XXXIX.

*Column Twenty-eighth, Cauftic Volatile Al-
kali.*

THE order of attractions for acids
feems to be the fame in this inftance
as that of the fixed alkalis; but the
volatile alkali takes up feveral metals,
which are left untouched by the others.
That

That zinc precipitates the other metals, when diffolved in volatile alkali, is certain, (Scheme 18.), and for the fame reafon as thofe which are diffolved in acids; a reafon that has been formerly mentioned, and will be farther illuftrated hereafter, (XLVII.); on this account I have placed them in the fame order as in the columns of the acids, till experiment makes us acquainted with a better.

THE influence of phlogifton on the folution of metals in acids, has been fhewn above in various inftances, (V. XIII. XV. XVII.); let me now confider fome combinations of volatile alkali. To cauftic volatile alkali let fome filings of copper be added; if the phial be quite full and be immediately ftopped, no folution will take place, but if a little fpace above the liquor be left to the air, or the phial remain open for a quarter of an hour, in a few days a

folution

folution will be obtained as colourlefs as water; but if the ftopple be taken away, it foon grows blue at the furface, and the fame tinge gradually pervades the whole mafs. A folution thus coloured may be foon procured in an open phial. The folution lofes its colour, if new filings be added, and the phial be kept clofed for twenty-four hours. Thefe remarkable variations depend on phlogifton: copper is infoluble in its metallic form, but it is very foluble when a little of the inflammable principle is extricated. Such a privation is effected in a veffel, either open, or only filled in part, by means of the air which ftrongly attracts this principle, as will be fhewn hereafter, (XLVII.), and the effects will vary in proportion to the quantity. If juft as much is carried off, as is neceffary to render the copper at all foluble, the folution will be colourlefs; but if it be farther deprived of phlogifton, the calcined copper

yields

yields a blue folution, which is obvious
even to the eye, for the colourlefs folution is foon deprived of phlogifton,
by being expofed to the air, and therefore immediately begins to turn blue,
upon coming into contact with the air.
On the other hand, the blue folution
lofes its colour, in the way above mentioned, fince the alkali more willingly
attacks copper but a little dephlogifticated, than that which is much
calcined. The force with which volatile alkali attracts dephlogifticated
copper, promotes the effect of the air.

VOLATILE alkali detonates with nitre, whence it manifeftly appears to contain phlogifton, and that as a proximate principle, which is feparated by
the calx of manganefe, quickfilver,
gold, and various other fubftances, in
confequence of a ftronger attraction;
and then an elaftic fluid, of a peculiar
nature, is obtained, which probably is

the

the other principle: I fay *probably,*
for it has not yet been reduced to volatile alkali by the addition of phlogifton. Moreover, in this decompofition it deferves to be remarked, that
no figns of vegetable alkali occur, fince
many contend that the volatile origi-
nates from the vegetable, intimately
combined with phlogifton.

In the dry way, the attractions of
volatile alkali feem to differ from thofe
immediately preceding in this, that it
may be expelled by fire alone, when
united with certain fixed matters;
and on this account I have thought
proper to exclude them entirely.

XL.

XL.

Column Twenty-ninth, Caustic Ponderous Earth.

THE celebrated Margraaf afferts that the bafe of ponderous fpar is calcareous earth, and indeed experiments fhew that they agree in various properties ; but Dr Gahn and Mr Scheele, by a more particular inveftigation, detected a great difference between them, and from my experiments it appeared to be ftill greater. Both agree in effervefcing in acids; in lofing the aerial acid, when expofed to fire, and thus acquiring folubility in water, and affording a cream in the open air ; in rendering alkalis cauftic, and in diffolving fulphur, &c. But they differ widely at the fame time ; for gypfum is lighter, and totally foluble in water, but ponderous earth forms with the vitriolic acid ponder-

ous

ous fpar, of which the fpecific gravity
is 4,500, and which can fcarce be at
all diffolved in water; with nitrous
and marine acids lime forms only
deliquefcent falts, but ponderous earth
difficultly foluble cryftals; lime faturated with acetous acid affords cryftals, but ponderous earth an almoft
deliquefcent falt: finally, they differ
widely in their attractions, as is evident from the preceding obfervations,
and will now be made to appear more
clearly.

THIS earth is but fparingly fcattered
over the furface of the earth, and has
hitherto been found only in combination with the vitriolic acid; a combination fo clofe that it can folely be deftroyed by the phlogifton of oils and
charcoal, with the affiftance of fire, and
then with difficulty. To obtain it pure,
it fhould be afterwards diffolved in nitrous acid, and precipitated by aerated
fixed

fixed alkali, which is effected by virtue
of a double elective attraction, for it
cannot be diflodged by cauftic fixed
alkali, (XIV.).

PONDEROUS earth forms, with moft
acids, falts of difficult folubility. From
all, the fmalleft drop of vitriolic acid
immediately throws down ponderous
fpar in the form of powder ; wherefore
I know nothing better than a folution
of this earth, in acetous or marine acid,
for detecting the fainteft traces of vitriolic acid, for it takes it from every
other bafis.

THE acid of fugar occupies the next
place. If it be dropped into a faturated folution of ponderous earth, in
the nitrous or marine acids, it feparates,
in a few minutes, pellucid cryftals, confifting of the earth combined with the
acid that was added. The fame acid
decompofes the compounds formed of

terra

terra ponderofa, and the acids of amber, of fluor, phofphorus, of the perlate, and that of fugar of milk, which
are all ftronger than the marine acid.

THE acid of amber comes into the
third ftation ; that of fluor into the
fourth, for it precipitates folutions
made by acids of amber, fluor, forrel,
phofphorus, of perlate falt, nitre or fea-
falt ; and left any one fhould afcribe
this to the prefence of vitriolic acid, let
it be obferved that this fediment, when
collected and added to vitriolic acid,
gives out fluor acid.

THE acid of forrel expels the phofphoric and perlate acids, that of fugar
of milk, nitre and fea-falt, and forms a
compound fcarce foluble, which come
next. The ftrength of the febaceous a-
cid has not yet been determined; in the
mean time I have placed it after the
marine. The remaining form this
feries : acid of lemon, tartar, arfenic,

ants,

ants, milk, benzoin, vinegar, of borax, of vitriol and nitre phlogifticated, the aerial, and that of Pruffian blue. The relative fituation of the firft mentioned individuals wants confirmation. Acid of arfenic does not vifibly reñder acetated ponderous earth turbid, but feems to form a triple compound; it, however, palpably expels the acid of benzoin.

I CONJECTURE, that, in the dry way, our earth comes next to the fixed alkalis; but I exclude the other earths, fince the ponderous fcarce fufes with them. Befides, I have fubjoined fixed alkali, and calx of lead, fince it enters into fufion with thefe.

XLI.

XLI.

Column Thirtieth, Lime

Pure calcareous earth, or lime, in the ftrict fenfe of the word, is that which is free from every heterogeneous fub- ftance; but by this appellation I indi- cate the abfence of acids, efpecially the aerial. Its attractions are very dif- ferent from thofe of ponderous earth. The firft place belongs to acid of fugar, which takes lime from every other, (Scheme 14.). To this fucceeds acid of forrel, which decompofes even gyp- fum, by attracting its bafis. Vitriolic acid exceeds the nitrous, and the others, (Scheme 16.). Acid of tartar takes lime from that of amber, the phof- phoric, perlate, and the following; and in like manner phofphoric acid, the perlate, and that of fugar of milk, from the nitrous, marine, fluor, arfenical, of

ants,

ants, milk, lemon, and vinegar. Fluor
acid attracts lime more powerfully than
that of ants and vinegar ; but is fcarce
fuperior to the nitrous and marine,
unlefs by the aid of water, and a double
attraction. Acids of lemon and benzoin exceed vinegar. The arfenical
acid does not difturb formicated and
acetated lime, unlefs the folution be
concentrated. The place of the febaceous is ftill uncertain. The laft
individuals follow the order of the
preceding column.

WHAT is faid in the preceding paragraph may be applied to the attractions
of lime, magnefia, and clay, in the dry
way, except that the laft fcarce attacks
fulphur.

XLII.

XLII.

Column Thirty-firſt, Cauſtic Magneſia.

This ſalt, which is called in the ſhops *magneſia alba,* differs in various reſpects from lime, as I have ſhewn in a particular eſſay. I once thought, that this earth attracted acid of fluor moſt forcibly, and that all others were expelled by it, (Scheme 15.); but the repetition of the experiments gives room to ſuppoſe that I was miſled by the ſiliceous ſediment: I therefore aſſign the firſt place to acid of ſugar, and ſo on as in the column; of which the order is for the moſt part eſtabliſhed by experiments, in the diſſertation above mentioned. Some places are, however, uncertain, eſpecially thoſe occupied by the acids of ſorrel and lemon. Calcined magneſia is inſoluble in water, which

which is a great obstacle when we are endeavouring to ascertain attractions.

XLIII.

Column Thirty-second, Pure Clay.

UNDER this denomination I understand earth of alum well purified, for common clay is always more or less mixed with the filiceous in powder: it contains, however, the real basis of alum, and thence derives its peculiar properties.

THE vitriolic acid attracts clay more powerfully than any other, next the nitrous, and then the marine acid. The order of the rest is not sufficiently ascertained; it is, however, certain, that the acids of fluor, arsenic, sugar, tartar, and phosphorus, take clay from the acetous. The determination of the attractions is in this case prevented, not

only

only by the infolubility of pure clay,
as in the cafe of magnefia, but likewife
by the obfcure cryftallization or de-
liquefcence of the compounds, which
commonly conceal the decompofition.

XLIV.

Column Thirty-third, Siliceous Earth.

WHILST the vitriolic expels from
fluor its proper acid, the furface of
the water in the receiver, even in the
gentleft heat, is gradually covered with
a white powder, which foon forms a
cruft. When this is broken, and finks,
another is generated, and fo on, as
long as any acid of fluor paffes over.
This matter, when collected and wafhed, has all the properties of filiceous
earth. I collected a parcel, and fent
it to Mr Macquer, who affured me
that it had all the properties which filiceous

liceous earth ſhews in the focus of a
burning glaſs. Now, whence comes
this powder? Does it exiſt in the fluor,
and is it volatilized by the heat? Or is
it extracted from the glaſs? Or is it
formed from its principles?

I by no means deny, that fluor ſometimes contains ſiliceous particles, but
they are accidentally preſent; for that
of Garpenberg, which I generally uſed
in this operation, ſometimes contains
not a particle. This, when reduced
to the fineſt powder, is totally ſoluble,
by a long digeſtion, in aqua regia,
which would never happen if it contained flint. Hence, therefore, it is
evident, that ſiliceous earth enters the
compoſition accidentally, and that the
powder collected in the receiver is by
no means to be aſcribed to it, ſince
no ſuch ſubſtance exiſts in the compound. But perhaps other ſpecimens
from Garpenberg are more or leſs mixed

ed with it. Glafs indeed abounds in filiceous earth, and is corroded during the procefs; but as Mr Scheele faw the powder appear when metallic veffels were ufed, I have long thought that in this experiment there was a generation of filex. Into a phial of iron or copper, upon which concentrated vitriolic does not act, there was put fluor mineral in powder, with an equal weight of ftrong vitriolic acid; the cover was then applied, to which were faftened below feveral different bodies, fome dry, and others wet. The phial was expofed for feveral hours to a gentle heat of digeftion, and upon opening it all the moift fubftances were found to be covered with powder, and the dry ones quite free from it. But the fluor ufed in this cafe, without doubt, contained filiceous earth; for Mr Meyer has fince found, by accurate and well conducted experiments, that no filiceous powder appears in metallic veffels when

when no glafs is ufed. The fluor with
vitriolic acid yielded nothing; but the
fame mixture, in equal quantity, fet in
a metallic veffel, when glafs was added
to it, fwelled very much, and a filiceous powder was volatilized. The fluor
acid, therefore, when refolved into vapour, feizes filiceous earth with violence, extracts and retains it till it is
abforbed by water; for upon entering
into this new combination, it is forced
to depofit part of the filex : the reft
remains diffolved in the acid liquor,
and may be thrown down by alkaline
falts *. But no other acid is capable
of diffolving filiceous earth, not even
in the very tender ftate of a precipitate,
from liquor filicum †.

FIXED alkalis, efpecially when cauftic,
diffolve very fine filiceous powder in
the

---

* Opufc. vol. ii. p. 34.

† *Ibid.* p. 36. and vol. iii. p. 314.

the moift way, but far more plentifully in the dry. All the acids, not even the aerial excepted, effect a feparation.

Liquor filicum is precipitated even by fluor acid; but the powder which falls does not confift of pure filiceous earth, for it contains likewife fluorated alkali, and thus exhibits a triple falt.

Water entirely rejects filex in a moderate temperature. I always procure by evaporation a fmall portion of filiceous earth from the water of the Upfal fprings, notwithftanding it has paffed through the filter feveral times; but it without doubt is fo fine, as when once mixed with the water, to be retained by friction; for divifion increafes the furface, and with it the friction, which at laft becomes equivalent to the difference of fpecific gravity.

In the dry way, filiceous earth is fufible with borax, minium, and other
proper

proper fubftances, but moſt eafily by fixed alkali.

SOME of the moderns reckon *earth of ivory* among the fimple earths, but improperly, for it is doubtlefs a compound, and, like earth of hartſhorn, contains both aerated and phofphorated lime.

XLV.

Column Thirty-fourth, Water.

SINCE fand in a very attenuated and volatile ſtate preferves a level, ſo that a-nimals may be drowned in it ; ſince pounded gypfum fet in a kettle over the fire feems liquid, not to mention other inftances, why may not liquids in general be confidered as folid molecules too fubtile to be perceived by the fight, however aſſiſted, and on account of their levity, bulk, figure, or by the interpofition of another fluid, moveable

with

with the greateſt eaſe, and affecting an horizontal ſurface? Hence water would ſeem to be nothing but earth kept liquid by heat. It certainly concretes into a ſolid body when the heat is diminiſhed to a determinate degree; but when the heat is increaſed to a certain term, it is reſolved into elaſtic vapours.

Every particle having a certain force of attraction for the principle of heat, forms a little atmoſphere for itſelf. As long as theſe atmoſpheres prevent the particles from coming into contact, the whole remains liquid; when they are enlarged, the diſtances increaſe, and an expanſion takes place, till at laſt the ſuperficial particles are reſolved, by the neceſſary quantity of heat, into elaſtic vapour. At 212° the whole maſs undergoes this change. The vapours then ariſe in great abundance, and produce the agitation of boiling. They exceed the bulk of the water 14,000 times, and

and then the vaft furface is able to abforb a far greater quantity of heat than
before ; and is it not thus that all evaporation produces cold ? On the other
hand, the excefs of heat being gradually diminifhed by the coolnefs of the
air, or in any other way, the bulk of
the vapours is contracted, and they are
condenfed at laft into drops of water.
If the matter of heat goes on to decreafe, the particles drawing near to
contact lofe their refpective mobility,
and concrete into ice. What is hereafter to be faid on the fubject of fpecific heat, will farther illuftrate thefe remarks, (XLVIII.).

Saline, gummous, and fpirituous fubftances are efpecially foluble in water.
In what order the falts are taken up,
has been hitherto little examined, nor
is this an eafy tafk. Concentrated vitriolic acid takes water from a folution
of vitriolated vegetable alkali, of alum.
                              vitriol.

vitriol, corrofive fublimate, and other fubftances, which it does not decompofe, fo that they cryftallize almoft inftantaneoufly. The other acids fcarce exert this power perceptibly.

CAUSTIC alkalis likewife attract water ftrongly, and precipitate various falts which they do not decompofe.

WATER attracts fpirit of wine more forcibly than thofe falts which are infoluble in fpirit of wine, and which therefore may be precipitated by it. This is the cafe with volatile alkali, which, when thus precipitated, is called *Van Helmont's foap, Sapo Helmontii.*

THE elective power of water with refpect to neutral and middle falts is hitherto unknown, and has been totally neglected. It is, however, probable, that each is attracted with unequal force, and that one gives way to another.

ÆTHER

Æther may in some measure be separated from spirit of wine by water.

XLVI.

Column Thirty-fifth, Vital Air.

A.] The atmosphere which every where surrounds our globe, consists of a pellucid, elastic, and apparently homogeneous fluid, which we denote by the name of *common* or *atmospheric* air: when more closely examined, it is found to contain, besides vapours which vary wonderfully, according to the diversity of situations and winds, three fluids mixed together, and widely differing in their nature. The greatest part, which certainly exceeds the others three times or more in bulk, is neither fit for supporting fire, nor for respiration, probably derives its origin from the vital

part,

part, in confequence of fome change
not yet certainly known, perhaps proceeding from the addition or fubtraction of phlogifton, and may therefore
be called *corrupted* air. That which
in England is called *dephlogifticated*,
and which I have formerly termed uncombined, good or pure, but now think
with the hiftorian of the Parifian Academy, that it fhould be diftinguifhed
by the appellation of *vital;* this, I fay,
differs widely from the preceding, being not only fit, but indifpenfably neceffary for fire and refpiration, every
other air being mephitic. The aerial
acid forms the fmalleft part of the atmofphere, fcarce ever amounting to
one-fixteenth.

B.] Vital air is found but fparingly
mixed with the atmofphere, amounting
fcarce to one-fourth of its bulk, feldom
or never exceeding one-third, as I have
juft intimated. It may, however, be
obtained

obtained by various means. An ounce
of nitre, expofed to heat in a pneumatic apparatus, affords 500—600 cubic
inches of air, far better, efpecially at
firft, than common air. Nitrous acid
poured upon many metals and earths,
and then abftracted to drynefs, yields
firft nitrous air, if any phlogifton be prefent, and then more or lefs of vital air.
Vitriol of iron, copper, and zinc, and
various vitriolated earths, nay, lapis ca-
laminaris, manganefe, and the calces of
the noble metals acquired by precipitation, afford, when expofed to a due degree of heat, a portion of vital air during their reduction. Hence it may
be juftly concluded, that this air can
indeed be obtained without nitrous
acid, but that by means of it a much
larger quantity is procured, fo that it
is fcarce to be doubted but that it enters as a principle into the acid, or the
acid into it. In the former fuppofition, fomething muft be removed, which
when

when again added to the vital air in
proper quantity, muft impart to it, befides other properties, a ftrong acidity.
The nature of this is unknown. Nitrous air alone is at leaft infufficient,
unlefs the exiftence of phlogifton be
denied, which, however, would be contrary to evident experiments ; nor
does it feem to be the inflammable
principle, by which, on other occafions,
all the acids are debilitated, and, when
fully faturated, are as it were fettered.
It fhould, however, be remembered,
that compounds fometimes have properties not belonging to either of the
ingredients. But in the cafe in queftion, this bare poffibility is unfupported by any experiment, which fhews
that diftinct acidity proceeds from
phlogifton. The latter propofition is
countenanced by the following confiderations. It appears from experiment, that nitrous acid very forcibly
attracts phlogifton, is rendered much

weaker

weaker by it, and at length, by the aid
of a certain portion of the matter of
heat, acquires an aerial form, being
changed into the aeriform nitrous acid:
with a greater portion of phlogifton,
nitrous air is formed, containing the
acid faturated; may not, therefore, a
farther addition of phlogifton or fpecific fire, as the increafe of weight feems
to indicate, or both thefe caufes, produce a new variation, and may not this
be vital air? This opinion is, however,
liable to difficulties, which fhall be
mentioned in the fequel. Meanwhile,
granting thefe pofitions, and affuming
Mr Scheele's hypothefis, the phænomena accompanying its generation admit
of a confiftent explanation. We know
that the acid as prefent in nitre may be
phlogifticated by fufion. The cohefion
is diminifhed in proportion to the increafe of phlogifton, and by a certain
quantity is totally deftroyed, fo that it
is changed into vital air, and may be
expelled

expelled by heat. The acceffion of phlogifton is derived from the decompofition of heat; now, by the feparation of one principle, the other which is left uncombined, according to the hypothefis, is vital air, and which therefore comes from two fources, compofition and decompofition. If thefe things be true, we can eafily comprehend how it is obtained from nitrous acid, poured upon almoft any fubftance; and alfo how we come to procure a new portion, by adding frefh nitrous acid, when the former portion has been exhaufted. So great is the volatility of this acid, that it efcapes before it is fufficiently concentrated to receive phlogifton; the bafis fixes the acid, and when it contains phlogifton, contributes, at the beginning of the procefs, while the heat is moderate, to the production of nitrous air.

THE

THE fubflances which furnifh vital
air without nitrous acid, contain, for
the moft part, fome metallic principle,
which, when acted upon by a ftrong
fire, decompofes either their inherent
fpecific heat, or that which flows in
through the veffels. The noble calces
feem by thefe means to be reduced
to a metallic form, without the addition of phlogifton, even fuch as neither
receive any aerial acid, nor contain
any nitrous air, as gold diffolved in
dephlogifticated muriatic acid, and afterwards precipitated. The ignoble
calces of lead, iron, zinc, manganefe,
and perhaps others, are alfo capable
of decompofing heat, but cannot retain
a quantity fufficient for their reduction. Manganefe does not indeed
grow black from the action of heat a-
lone ; but by the affiftance of vitriolic
acid, it fixes a quantity fufficient to
render it foluble. Such phænomena
fometimes, though rarely, occur in the
humid

humid way. Upon minium dephlogifticated, by a long continued calcination, and put into a fmall retort, let a
little more than its bulk of concentrated vitriolic acid be poured; the mixture
foon grows warm, and acquires a black
colour, a phænomenon which is produced by heat alone; then fome degree
of effervefcence takes place, during
which, fome aerial acid is extricated;
but in a minute or two the heat becomes infupportable to the touch, and
at the fame time there iffues out with
vehemence a white fmoke, which yields
pure vital air: the volatilization of the
powder gives it at firft the appearance
of fmoke. The addition of pulverized
glafs renders minium more penetrable
to the vitriolic acid; without this addition, the upper ftratum of the mafs
grows white, while the lower ftrata ftill
remain black, or are not yet acted upon
by the acid; in which cafe, aerial acid
is extricated during the whole time of
the

the production of any elaftic fluid. I
have treated minium in like manner
with the muriatic acid; but in this o-
peration the menftruum muft be made
to boil. Then aerial acid is extricated, as alfo dephlogifticated marine a-
cid, and the vital air fcarcely amounts
to one-fourth. I have no doubt of the
exiftence of aerial acid in the minium;
but it remains to be tried, whether it
be prefent in minium recently prepared. I cannot determine this, fince it
is not made in Sweden. It probably
does not unite with the lead during
calcination, but is afterwards attracted
from the atmofphere, as happens with
regard to lime. But the origin of the
vital air is an important enquiry.
Some deduce it from the dephlogiftication of the aerial acid; Mr Scheele
from the decompofition of heat. All
minium, hitherto examined, contains a
fmall portion of magnefia nigra, which
is capable of decompofing the matter
of

of heat, and lead may perhaps have the
fame power. Muriatic acid is more
eafily decompofed ; and therefore, by
employing this menftruum, we gain
but little vital air. Aerated lead,
treated in the fame manner, yields only aerial acid ; but nitrated mercury,
calcined to rednefs, and the fpontaneous fediment of vitriolated iron, when
treated with vitriolic acid, produce
nearly the fame effects as minium.
Iron is fcarce ever without manganefe,
and feems itfelf capable of decompofing heat ; for that which is not magnetic, acquires this property in the
crucible without any addition of phlogifton.

THE great fpecific heat of vital air,
which can be fhewn by proper experiments, renders its origin from nitrous
acid, by the addition of phlogifton, not
a little fufpicious ; and perhaps future
experiments

experiments will fhew that it is not the real one.

C.] The relation of vital air to other bodies is now to be confidered. There is no fubftance hitherto known, on which vital air acts more readily and efficacioufly, than on nitrous air. At the inftant of contact, the whole mafs almoft hiffes, turns red, grows warm, and contracts in bulk. The enquirers into nature are not yet agreed about the caufe of thefe phænomena. There are two prevailing opinions, of which one will probably prove true. Mr Kirwan has thus explained it *. Phlo-gifton is more ftrongly attracted by vital air than by nitrous acid ; where-fore the nitrous air is dephlogifticated, lofes its nature, and the red vapour of nitrous acid is produced, which water readily abforbs. But vital air, faturated with phlogifton, forms aerial acid, and

* Phil. Tranf. 1782.

and with a ftill larger portion is converted into corrupted air. The bulk is fometimes diminifhed to $\frac{3}{100}$, in confequence of the deftruction of the nitrous air, and the generation of aerial acid, which being heavier than the vital air, and having lefs fpecific heat, ought neceffarily be contracted, and, moreover, may be totally abforbed by water. The incalefcence is produced by the deftruction of the nitrous air, which, as well as the phlogifticated vital air, parts with a portion of fpecific heat. This is a very ingenious explanation; but before it can be received as quite fatisfactory, fo much aerial acid, as it fuppofes, muft be more precifely demonftrated. If I miftake not, the production of aerial acid is deduced from the precipitation of lime-water, and its quantity from the diminution of bulk. With refpect to the former conclufion, it is well known that lime-water is manifeftly rendered

turbid

turbid by a very fmall quantity of
aerial acid, accidentally mixed with
vital air ; but the diminution, if it be
repeatedly paffed through recent lime-
water, may arife from another caufe,
for water abforbs one-fourteenth of its
own bulk of vital air ; if, therefore, a
fufficient quantity of water be prefent,
the whole diminution may happen in
this manner. In Mr Scheele's hypothefis, the relation of vital to nitrous
air is alfo made to confift in the dephlogiftication of the former ; but it is
faid, that the matter of heat is generated by this combination, which, in
the prefent cafe, is not abforbed by any
of the furrounding bodies ; for the nitrous air changing from fluidity to a
liquid ftate, muft give out a portion
of fpecific heat, to be diftributed a-
mong the contiguous fubftances, and e-
vidently paffing through glafs. On what
foundation this origin of heat refts, I
fhall hereafter examine, (XLVIII.).

THE

THE inflammable air of metals contains phlogiston almoſt pure, in an aerial form ; yet it neither changes vital air, nor is changed by it, not even when the ſurface of contact is increaſed by the acceſſion of external heat ; a circumſtance which ſeems ill to agree with what has been juſt ſaid of nitrous air. But it is to be obſerved, that the decompoſition of nitrous air is the effect of a double attraction ; the phlogiſton is attracted by vital air, and the acid part by water. Therefore, when the mixture is made in a phial immerſed in mercury, the experiment fails. But the contact of flame, or a glowing body, produces a wonderful effect in the mixture of inflammable with vital air ; for it takes fire with ſo much violence, as far to exceed common fire in heat and efficacy, and to dazzle the eyes with the brightneſs of the light. If the conflagration be performed in a ſpace cloſed by mercury, as may be done by

prudent

prudent management, the whole bulk of air is found to have in great meafure difappeared after the extinction of the flame, and the cooling of the apparatus. The refiduum affords corrupted air, fcarce ever mixed with aerial acid. This diminution is explained, as before, in two ways. It is impoffible to deny, that a contraction of bulk muft take place, if vital air be changed, by combination with phlogifton, into aerial acid, which is heavier, and has lefs fpecific heat; but the farther change of aerial acid by phlogifton into corrupted air, a lighter fubftance, and therefore neceffarily occupying a larger fpace, feems ill to agree with the fmall portion which remains after deflagration, though even the whole of the inflammable air fhould become fixed.

If we fuppofe both airs perfectly pure, this inflammation would feem the moft fimple of all; for there is no fuperfluous principle prefent, either to

give

give out any extraneous fluid during the operation, or to yield any refiduum which greedily abforbs a portion of the mixture.

ELECTRIC fparks feem to come near-eft to this operation ; they may be con-fidered as fmall flames changing the vital air which they meet with on the furrounding medium, by means of their phlogifton, either into aerial acid or heat, as muft be determined by future experiments. Alkaline air is changed by the electrical fpark into inflammable air of thrice its bulk, which fufficient-ly fhews that thefe fparks give out phlogifton in a free and elaftic ftate.

IN vital air, without the aid of ex-ternal heat, phofphorus is confumed very flowly, and fcarce at all, unlefs water be prefent to forward the decompofi-tion, by a double attraction. It may be burned by proper management in a glafs veffel clofed with mercury.
The

The method that beſt ſucceeded with
me, was to introduce a ſmall piece into
the glaſs veſſel, and ſet fire to it, by applying the flame of a candle externally.
This was repeated as long as the bits
ſucceſſively introduced could be made
to deflagrate.   After the apparatus had
grown cool, there never remained above
one-fourth, often not above one-tenth
of air, and ſometimes ſtill leſs ; the
reſiduum conſiſted of corrupted air,
very ſeldom mixed with aerial acid.
Here the hypotheſis, concerning the
origin of the aerial acid and corrupted
air, from the phlogiſtication of the vital, ſeems ſcarce admiſſible.   We have
in this experiment no uncombined phlogiſton, by conſolidation of which the
bulk might be diminiſhed.   The difficulty is, however, leſſened, by the ab-
ſorbing power of the uncombined phoſphoric acid, of which more will be
ſaid hereafter.   Sulphur may be burned

in

in the fame way, and exhibits almoft the fame phænomena.

THAT kind of inflammation is moft complex, which confumes a body that gives out aerial acid from its own fub-ftance, and at the fame time yields an ab-forbent refiduum. To this head belongs the combuftion of pyrophorus, candles, and other animal and vegetable fub-ftances. There can be no inflammation without vital air, whence it may be confidered as the pabulum of fire; yet the phænomena vary according to the nature of the bodies to be confumed. It is afferted, that in a fpace limited by mercury, a candle which is left to burn as long as it can, diminifhes the air lit-tle or none; and it is certain that the bulk is found the fame after cooling, for the confumed fat yields a quantity of fluid nearly equal to the vital air con-fumed by the flame. It is known that
oils,

oils, when decompofed in clofe veffels by fire, gives out aerial acid.

HEPATIC air, dephlogifticated by vital, depofits fulphur but flowly ; whereas nitrous air is decompofed at the moment of contact.

METALS, being loaded with phlogifton, cannot but be expofed to the power of vital air. The ignoble ones are decompofed more or lefs quickly, according to circumftances ; the noble refift obftinately : however, the pureft gold, when fufed, and fufficiently rarefied by the focus of a lens, is forced to part with fome of its phlogifton *. Mercury, which feems intermediate between the noble and the ignoble, when fufficiently heated, is well and quickly calcined in vital air, but remains unchanged in corrupted air, as Dr Prieftley has found. I have experienced the fame

* Macquer Dict. de Chimie.

fame thing with that mixture of lead,
tin, and bifmuth, which is fufible in the
heat of boiling water. The effects are
here alfo explicable from the generation of aerial acid, or the matter of heat.
On the former fuppofition, a difficulty
arifes from the rejection of aerial acid
by many metallic calces, and from the
reduction of gold without addition,
when it has been diffolved in dephlogifticated marine acid, and precipitated
by alkali, when the calx is neither contaminated with aerial acid nor nitrous
air. By Mr Scheele's fyftem, thefe difficulties are avoided, but others occur
not eafily to be obviated by the decompofition of heat. The calcination of
lead by mercury in common air *,
which is diminifhed one-fourth, and
yields an aerated calx, is among them ;
for I have never found in common air
a quantity of aerial acid amounting to
one-fourth of common air, nor, I believe,

* Mr Kirwan.

lieve, has any one elfe: it remains,
however, to be tried, whether the bulk
of the aerial acid, when expelled, cor-
refponds to the above-mentioned diminution. The increafe of weight is
afcribed, with great probability, to aerial acid, when it is prefent; but I can
fcarce doubt that fomething is contri-
buted by the increafe of fpecific heat.
Nor is the abforption of moifture by a
fpongy mafs, like that in queftion, always to be neglected.

BESIDES thefe fubftances, many more
are undoubtedly changed by the action
of vital air, efpecially thofe containing phlogifton; but the mode is unknown.

XLVII.

XLVII.

Column Thirty-sixth, Phlogiston.

THIS very fubtile matter admits, like the aerial acid, of two ftates, a ftate of combination, and freedom. In the former, it enters into the ftructure of bodies, eludes all our fenfes, and can only be recognifed by its effects, for which reafon fome have fuppofed it to be a fictitious fubftance, and totally impalpable, but without juft reafon. The two celebrated philofophers, Prieftley and Kirwan, have clearly proved its exiftence, both analytically and fynthetically, fo that I think all reafons for doubting are now removed. This principle, when in combination, and then it is properly called phlogifton, may be fet loofe by various methods; having recovered its elafticity, and

gained

gained an aerial form, by a proper increafe of fpecific heat, it receives the
name of inflammable air. In the next
paragraph, we fhall find data from the
analyfis of charcoal for eftimating the
weight of phlogifton in inflammable
air; a cubic decimal inch of inflam-
mable air is equal in weight to $\frac{63}{1000}$ of
an affay pound, and it contains as much
phlogifton as two pounds of forged iron,
i. e. $\frac{5}{100}$ *; therefore $\frac{63-50}{1000}=\frac{13}{1000}$ give
the weight of fpecific fire neceffary to
the aerial form, of which more will be
faid in the next paragraph, (XLVIII.
C. E.). I fpeak here only of the inflammable air of metals : that which
organic bodies yield, appears to be lefs
pure, and efpecially combined intimately with a portion of aerial acid.

PHLOGISTON is perhaps to be found
in all bodies, though in many it is concealed by its exility. The attractions
of

---

* Analyfis Ferri, p. 24.

of the more remarkable combinations
into which it enters can alone be examined here; a taſk which is incumber-
ed by no trivial obſtacles. Magneſia
nigra, for inſtance, attracts it with ſuch
violence as to decompoſe acid of ſalt.
It takes this principle from all the metals, but not without the aid of ſome
acid ; a circumſtance to be carefully noted. Nor does it act with ſo great force,
till it has obtained the quantity neceſſary to perfect ſaturation, but only till it
has acquired that which is neceſſary to
its ſolubility in acids. When this point
has been once attained, the complement which effects complete reduction
is attracted more feebly than by any
other metallic calx. An attraction of
this kind, ſtopping at a certain point,
takes place in many other metals, though
it has hitherto been little examined.
Thus, calx of iron, and perhaps of the
other ignoble metals, by expoſure to
heat, acquires phlogiſton enough to become

come magnetic, but cannot acquire enough for reduction. In general, the reducing portion of phlogifton adheres much more weakly than the coagulating. Thefe attractions are in fome meafure analogous to thofe which acid of tartar exerts upon falts containing vegetable alkali, (XXXVII.). When the doctrine of affinities is brought to perfection, I forefee that it will often be neceffary to adduce the fame fubftance in two or more different ftates. Here the black and white calx of manganefe may be introduced feparately; but the former does not act by its fingle power, for it requires to be affifted by an acid. That, therefore, which is placed in this column is the white.

A PLACE can fcarce be allotted to vital air, as it has fcarce any effect, unlefs it be affifted by a double affinity, or a great degree of heat. We have before

fpoken

fpoken of its action on nitrous acid. Either external heat, or furrounding moifture, is neceffary to the complete decompofition of phofphorus.

CALCINED mercury is reduced by digeftion in acid of falt; but the caufe has not yet been fufficiently explored. As this acid, when dephlogifticated, attacks the metal itfelf, the calx can fcarce dephlogifticate the acid. It remains, therefore, to be examined, whether dephlogifticated air is produced during digeftion. According to Mr Scheele's hypothefis, the decompofition of heat is fufficient.

VOLATILE alkali is dephlogifticated by magnefia nigra; but the caufe is complex, depending upon nitrous acid.

THEREFORE thefe phænomena are at prefent of no ufe in determining the elective attractions; but the following
are

are more fimple, and feem adapted to
this end.

NITROUS acid decompofes fulphur,
very flowly indeed without boiling ; but
it feparates the principles of muriatic
acid in a middle temperature.

DEPHLOGISTICATED marine acid does
not act upon fulphur ; but it gradually
decompofes white arfenic, and immediately refolves phofphorus into a
white fmoke.

THOUGH the precipitation of metallic folutions by complete metals is really the confequence of a double attraction, yet a fingle attraction would be
fufficient, could phlogifton be fupplied
in a proper ftate. The inflammable
principle has a different attraction for
different calces, and combines with
them to faturation ; after which, the
metals fall down in a complete ftate,
and

and cannot be rediffolved, unlefs the excefs of phlogifton be removed. When any other metal whatever is put into a folution of gold, the gold is immediately precipitated, not on account of the inferior attraction of the acid, as it has hitherto been univerfally explained, but becaufe phlogifton more readily unites with the calx of gold, than with the calx of the added metal. That this is the true caufe, may be fhewn both by the dry and humid way. With refpect to the latter, a fine difcovery, made by Mr Sage, throws great light on the queftion. He puts into a diluted folution of a metal a piece of phofphorus, which yields its phlogifton to the metallic calx, and in fome cafes completely reduces it. The calces of the noble metals, and of copper, thus recover their metallic ftate*. Though there is here no reciprocal exchange of principles, yet two powers

effect

* Rozier, Journal de Phyfique, 1781.

effect the decompofition. Water a-
lone gradually extracts the acid part
of phofphorus, but very flowly, and
thus renders the combination of the
metallic calx with phlogifton more
eafy. We have experiments yet more
direct. A folution of acid of arfenic
in water is made to acquire a reguline
form, by paffing a ftream of inflammable air through it, as Mr Pelletier
attefts*. The fame thing happens to
fome other metallic folutions. Metallic calces may be reduced by the flame
of pure inflammable air; which alfo
happens if they are immerfed in this
air in clofe veffels, and expofed to the
focus of a burning glafs, as Dr Prieftley
has found. The air is diminifhed in
this operation; but the refiduum retains
its former nature, and is juft as fit
for contributing to reduction as before. Befides, it is well known, that
fome metallic calces may be reduced
by

* Journal de Phyfique, 1782.

by fufion, with the addition of iron or any other proper metal.

THE phænomena, therefore, which come under this head, when deduced from their real caufe, totally invert the feries of metallic calces laid down in former tables. Thus, gold rifes from the laft place to the firft or fecond, and zinc is reduced to the loweft. This holds with refpect to the reft, as appears in the table of attractions.

IT is certain that the arfenical acid attracts the inflammable principle with greater force than the phofphoric; for if phofphorus be put into arfenical acid, the furface foon grows black in confequence of reduction.

IN the dry way, I have placed the metallic calces according to the order juft eftablifhed. I have placed the acid of arfenic before the calx of filver; for
this

this acid expofed to the action of fire
with filver, diffolves a portion of it,
which cannot be done without dephlogiftication. The dephlogifticating portion is fublimed in the form of white
arfenic ; the other immediately diffolves the calx.

Supposing the matter of heat to
confift of phlogifton and vital air, the
place of vital air is between the calx of
mercury, which is reduced as well as
the noble calces, and the ignoble calces. It cannot, however, be denied,
but that thefe alfo are capable of decompofing part of the heat, though not
fo as to effect a complete reduction.

XLVIII.

XLVIII.

Column Thirty-seventh, the Matter of Heat.

A.] The nature of fire exercised the genius of philofophers in the earlieft times, nor has the diverfity of opinion yet been reconciled. Nay, it has been made a queftion, *Whether the phænomena which are afcribed to fire are to be deduced from a peculiar matter?* Or, *Whether they depend only on the motion of the particles of bodies?* Now fince all motion, which is excited on our globe, meets with refiftance, and, therefore, when left to itfelf, is progreffively diminifhed, as every day's experience teftifies, it is not eafy to conceive how the motion excited by the produdion of a fpark, which muft meet with continual retardation, fhould neverthelefs fometimes acquire fuch augmentation as to be able to confume a houfe, nay, a
whole

whole city.   Here the effect far exceeds the cause.   But in this age, almoft all philofophers agree, that there
is a peculiar matter of fire, which has
gravity ;   exerts an attractive power ;
poffeffes other peculiar properties very
palpable in various cafes, and capable
of being accurately determined.   I
therefore think it fuperfluous to dwell
any longer upon the proof of this pofition.   The nature of this matter is a
point much more difficult to be determined, and affords a fine field for the
exertion of the greateft abilities.   I
think there can be no doubt that it
ought to be called the matter of heat
rather than of fire.   Fire is the action
of heat when increafed to a certain degree, and, therefore, foon paffes away
after the confumption of the fuel ; but
the heat continues though it becomes
rarefied, and is diftributed among other
bodies.   There is always heat in fire ;
but all heat is not fufficient for ex-
                                    citing

citing fire : a determinate accumulation is required in every cafe.

UNLESS, therefore, we fhould chufe
to invert the ufual mode of fpeaking,
the denomination which I have placed
at the head of the paragraph feems
more fuitable to the nature of the
thing.

B.] The chief opinions now prevailing concerning the matter of heat may
be referred to three fyftems.

Firft, Some confider light itfelf as
elementary fire, which every where
furrounds our planet, in an uncombined ftate, becoming lucid when in
fufficient motion, and occafioning different temperatures by its unequal denfity, highly elaftic, light, fubtile, and penetrating. Notwithftanding its wonderful tenuity and mobility, it may be
fixed in bodies, and enter into their
compofition

compofition as a proximate principle ;
in which ftate it is denominated phlogifton. The great fimplicity of this hypothefis recommends it ; but it can
fcarce maintain its ground, fince it has
been fhewn that uncombined phlogifton is nothing but inflammable air,
(XLVII.). Light feems moreover to
be inferior in tenuity to heat.

Secondly, Others argue, that elementary fire, which in a ftate of liberty occafions warmth, is not only different from
phlogifton, but fo oppofite that one
every where expels the other, at leaft
in part. Air during phlogiftication
gives out much fpecific fire, which,
when free, heats, calcines, caufes ignition, &c. It is proved, that the very
attenuated matter of heat is not equally diftributed, and in proportion to the
bulk of bodies, as Boerhaave affirmed,
but that each body, by a peculiar attraction depending upon its nature,
acquires

acquires a greater or lefs quantity. If the heat marked by the thermometer is increafed or diminifhed in a place where there are feveral bodies of the fame weight, it is diftributed among them in proportion to their powers; and in like manner, in reftoring the equilibrium which is difturbed by a diminution, they exonerate themfelves in proportion to their powers. A body abforbs more heat in becoming liquid than it contains when folid; and there is need of a ftill greater portion to induce the ftate of vapour. Animals grow warm by refpiration, &c. not to mention other phænomena which will be related below.

PART of the fyftem concerning the increafe of latent or combined heat, when a folid becomes liquid, or a liquid is converted into a fluid, owes its rife to the illuftrious Black. It has fince been cultivated with fo much fuccefs,

cefs, both in England and Sweden, that it now feems to reſt on a fure foundation *. The function of refpiration has been particularly illuſtrated by Prieſtley and Crawford.

THE *third* fyftem is that of my fagacious friend Mr Scheele, who thinks that the matter of heat is not fimple, but compounded of phlogifton and vital air, clofely combined, and that light confifts of the matter of heat, with an excefs of phlogifton. His Treatife on Air and Fire will beft ſhew how he arrived at thefe conclufions. This hypothefis is not without its difficulties, which I every where mention ; it however feems to agree better with experiment than any other, and therefore I have often adapted my explanations to

it.

* See Crawford on Animal Heat, Magellan du Feu Elementaire, and Wilcke, in the Stockholm Tranfactions of 1773,—1781.

it. It is by no means neceffary in this hypothefis, that the contraction of the bulk of the air fhould always be afcribed to the heat paffing through the glafs. There is no reafon why it may not be abforbed on particular occafions.

C.] I acknowledge that the new doctrine concerning the diftribution of heat is well eftablifhed in many refpects; but as it is connected with attractions, it will be proper to explain with greater accuracy in what light I view it.

LET the heat which we can meafure by the thermometer be called *fenfible*, and that which is fo fixed by the attraction of bodies, that it cannot be indicated by the thermometer, *fpecific*. For the fake of comparifon, the fpecific heat of water is expreffed by unity,

to

to which the specific heat of other
bodies of equal weight, and the same
sensible temperature, is referred and
expressed in numbers *, which indicate
the proportion, but not the quantities.
Let us suppose two bodies, *A* and *B*, of
the same weight, whose specific heats
are as *a* to *b*; let the sensible heat in
the vicinity of the bodies be increased
by the quantity *m*, which is to be divided between *A* and *B*, the former
will receive an increase $= \frac{a}{a+b}m$, and
the latter $= \frac{b}{a+b}m$, so that $\frac{a}{a+b}m : \frac{b}{a+b}m$
$:: a : b :: a + \frac{a}{a+b}m \quad b + \frac{b}{a+b}m$. This
is also the case if *m* be supposed to be
negative, or to denote a diminution, for
in either case such a distribution will
take

---

* See Mr Magellan, who has given this theory an
elegant mathematical form. The method of determining the specific heats differs from that of Wilcke,
but the events agree ; a circumstance which not a little
confirms the truth of the doctrine.

take place, that the proportions of fixed heat fhall remain the fame.

BUT the fpecific heats do not follow
the proportions of fpecific gravity, nor
of bulk, but, if I miftake not, the compound ratio of the peculiar attraction,
and the furfaces. I do not mean the
mere external furface, but the internal
likewife : it is well known that there
is no body perfectly folid, nay, gold
itfelf, the heavieft of all fubftances
hitherto known, is perforated with invifible pores to the amount of one-half
of its bulk, as has been rationally conjectured by Newton. Hence we may
form fome judgment of the vacuities
of other bodies, fince they may be at
leaft relatively determined by their
fpecific gravity ; but this affiftance is
of no ufe in the prefent cafe. Though
every other figure affords a larger furface, yet let us affume fpherical pores
for the fake of fimplicity, and the va-
cuity

cuity of any body reduced to a sphere of the diameter 10, the internal surface will be as the square, that is, as $10 \times 10 = 100$. Suppose now this vacant space to be divided into ten equal spherules, of which let the diameter, to avoid fractions, be expressed by m, and the internal will be as $10\ m^2$. If it be divided into an hundred spherules, it will be as $100\ n^2$, and so we may go on as long as we please. Thus the internal surfaces increase as the size of the pores decrease, and, in the same proportion, the specific heats, if I am not mistaken. As the pores amount always to more than half the bulk, and in most inorganic bodies altogether elude the fight, however assisted, by their minuteness, the external surface may be neglected as infinitely small, and this perhaps holds concerning the peculiar force of attraction. The internal structure of bodies may indeed be truly compared to a sponge, though

the

the apertures cannot in general be perceived. Now heat penetrates into all the pores of bodies, and when fixed, furrounds the fmalleft atoms like an atmofphere, and adheres to them, deprived of its power of exciting warmth. The thicknefs of this ftratum is increafed or diminifhed according to circumftances. The following are a few inftances which feem to confirm my conjecture, for the nature of the thing forbids us to expect a rigorous demonftration. The particles of water, in the ftate of congelation, touching each other at a greater number of points, cohere in confequence of their attraction. But if a fufficient quantity of heat penetrates into ice, the particles are gradually feparated, regain their little atmofphere, and recover their mobility. In a ftronger heat the mafs is dilated by larger atmofpheres furrounding the particles, and their tenuity cannot but be increafed by them.

Laftly,

Laftly, in the boiling temperature, every particle is fo much dilated, as to occupy a fpace 14,000 times greater, and fuppofing the form to be fpherical, as is in fome meafure vifible, acquires a furface about 600 times more extenfive. By this expanfion, the contact and the attraction depending on it increafes exceedingly, fo that a remarkable degree of cold is produced in the contiguous bodies, by the quantity of heat neceffary to faturation being collected and fixed.

My opinion is alfo illuftrated by the following facts. Let a thermometer, with a void fpace above the liquor, and with the top clofe, be fupended in the receiver of an air-pump ; as foon as the air begins to be rarefied by the ftrokes of the pifton, the liquor of the thermometer will fink, as was firft obferved by Dr Cullen. The defcent is owing to the dilatation of the glafs, in confequence

fequence of the removal of the external
preffure, for if the point be broken before fufpenfion, the level of the liquor
will not be changed by the rarefaction
of the air. A vacuum does not therefore of itfelf produce cold. But if
the globe of the open thermometer be
moiftened, the liqour will defcend on
the rarefaction of the air. The caufe
is to be fought in the ambient air, for
its particles being expanded afford a
fpace wider, and better fitted for abforbing heat efpecially from the water
on the ball, which repairs its lofs from
the glafs, and is converted into vapour;
the glafs attracts heat from the mercury, which therefore contracts, till
the equilibrium be gradually reftored
from the neighbouring bodies. We
have then three places filled of the column, at the top of which ftands the
matter of heat; the firft is occupied by
air, the fecond by glafs, and the third
by the liquid metal. That vapours
transfer

transfer heat to rarefied air, I conclude
from their fudden condenfation into
drops. It is obvious, that evaporation
is much forwarded, in this cafe, by the
rarefaction of the air. The moifture
is diffolved by the heat which flows
out, and is, therefore, expanded into
vapours that are vifible, and productive
of cold, as is well known. The air,
however, can fcarce deprive the glafs
of its heat, without the expanfion of
the water into vapour, for when rarefied, it acts very flowly on the dry
globe of the thermometer; abundant
moifture acts more efficacioufly than
when it is in fmall quantity; nay, in general, the more volatile is the liquor
ufed, the lower does the mercury defcend. Therefore vitriolic æther, highly rectified fpirit of wine, cauftic volatile alkali, water and effential oils,
fhould, it would feem, be placed in the
feries after air.

I HAVE formerly fhewn that heat is abforbed during the folution of falts which acquire a far more extenfive furface, and that it is let loofe again by fudden cryftallization *. For the fame reafon, muriatic air eagerly attracts phlogifton, (XVI.), not to mention other proofs of efficacy heightened by an increafe of furface. Beyond the fphere of contact there is fcarce any attraction, and therefore the area is of more importance than the denfity. For if as much be fixed to a denfe, but fmall, as to a rare but extenfive furface, the elafticity of the matter to be fixed will be more or lefs reftrained below the equilibrium prevailing in the contiguous bodies, but the attractive power can fcarce fuftain fuch compreffion.

EVERY body has a determinate fpecific heat, which, however, appears,

from

* Opufc. vol. i.

from experiment, to vary with the
ftate of the body. In the folid ftate
it contains leaft, in the liquid more,
and the fluid, in which there is the
weakeft cohefion, has the moft fpecific fire. Within the limits of the
ftate of folidity, no variation has as yet
been obferved, though without doubt
the fpecific heat ought gradually to increafe, in proportion to the approxima-
tion to liquidity, and *vice verfa.* It can
fcarce be doubted that fuch variations
are perceptible in the ftate of fluidity.

Does this ftratum of fpecific heat,
which involves the fmalleft particles of
a body, in any way affect the weight
of the whole? Without doubt this fubtile matter has gravity, and when it is
fo fixed to the body by attraction, as
not to act upon the thermometer, it
ought to caufe an increafe of weight. In
folids, indeed, it conftitutes but an infinitely fmall augmentation, fo that the
weight

weight cannot be obferved without great difficulty ; but in fluids, in which it abounds more, and bears a greater proportion to the weight of the whole, it ought not to elude the accuracy of our inftruments. Some experiments of Mr Lavoifier afford hopes that it may be actually determined *. That very accurate chemift burned fulphur and phofphorus inclofed in common air by means of mercury ; and when the apparatus was grown cold, he found that the acids, when freed from their combination, twice or thrice exceeded the burned materials in weight. Now whence comes this increafe? Let the table at the end of this paragraph be confulted, and it will be found that the fpecific heat of fulphur, and concentrated vitriolic acid, is as 0, 183 : 0, 758, that is nearly as 1: 4. But if we confider, that the fpecific heat of vitriolic acid increafes along with the inherent water, and

* Mem. de l'Acad. de Paris, 1777.

and that our ufual concentrations leave
a confiderable quantity of fuperfluous
water, we fhall be obliged to own that
the fpecific heat of the vitriolic acid,
deprived of all extraneous water, ought
to be reduced to 3, and probably
ftill lower. Now Mr Lavoifier affirms,
that the increments of the acid refidua exactly anfwer to the weight of
the vital air loft during the operation ;
whence, he juftly concludes, that this
air has been abforbed by the acid. As
the fpecific heat of phofphorus has not
yet been determined, a like deduction
cannot be made with refpect to it.
Meanwhile, thefe experiments, inftitu-
ted with a very different view, feem
not a little favourable to Mr Scheele's
hypothefis. When the vitriolic acid is
fet at liberty by the combuftion of
fulphur, its fpecific heat ought to be increafed in the proportion of 3 : 1. This
increafe is found upon experiment to
have taken place. At the fame time,

an

an equal weight of vital air is loft, and is quickly abforbed, together with that portion of phlogifton, which, when combuftion is otherwife performed, generates vitriolic air ; for the acid in Mr Lavoifier's experiments, was without fmell. The like phænomena occur in the combuftion of pyrophorus, as the fame chemift has found. Does not then the matter of heat coincide with the combination of vital air and phlogifton in the prefent cafe ? It has not yet been proved, nor is it indeed probable, that the aerial acid is among the principles of the vitriolic ; it does not emit pure air, when combined with pure alkalis, unlefs perhaps in confequence of the decompofition of heat, but a large quantity of uncombined heat is extricated. Moreover, I think this experiment fhould be repeated, not only in mercury but in dry air, in order to try whether the vitriolic acid can be obtained dry. It is as yet doubtful, whe-

ther

ther the liquid which we find, does not originate in part from extraneous moifture. A very accurate weighing of the refiduum likewife, without any foreign additaments, may perhaps ferve to determine the abfolute weight of heat, which promifes the illuftration of many obfcurities. Laftly, we fee the prepofterous manner in which the gravity of heat has been hitherto fought ; metals, the heavieft of all bodies, have been ufed for this purpofe, though they were of all others the moft unfit.

THE laft ten years are remarkable, among other things, for the change of many fubftances into an aerial form. And it is certain, that the generation of elaftic fluids is highly worthy of examination. The experiments which have been hitherto made, feem to indicate,

1. THAT

1. THAT the *substances liable to this change are the more simple*, especially salts, both alkaline and acid. We have long known, that the vitriolic, nitrous, muriatic, fluor, and acetous acids, as well as all those of vegetable and animal origin, as also the vegetable and volatile alkali, may be brought to the state of air. These, retaining their acid or alkaline nature, are readily abforbed by water, and ought, therefore, to be collected in veffels full of mercury. The aerial acid is as yet of obfcure origin. Moreover, sulphur * may be resolved into hepatic air; nay, filiceous earth, with all its fixity, may be made to affume the form of air †. Many more bodies will, doubtlefs, hereafter, be brought to the fame ftate. Gold itfelf may be converted into the form of vapour by means of fire; but whether

* Opufc. vol. ii. pp. 340. 345,
† *Ibid.* vol. iii. p. 397.

whether it can put on that of air, cannot be determined *a priori.*

2. THAT *the principal cause of this
transformation is to be sought in the matter
of heat,* which fubtilizes fubftances, and
gives them elafticity, by loofening their
particles. No elaftic fluid is found
without a large portion of fpecific heat;
nay, phlogifton itfelf is refolvable by
thefe means into an elaftic form. The
matter of heat comes in fome proceffes, either from the fire itfelf, when,
for inftance, the aerial acid is expelled
from chalk, inflammable air from iron
by heat alone ; or when it is extricated
by the vitriolic acid, for in the new
combination the acid cannot retain all
its fpecific heat ; what is fuperfluous is
extricated, and either totally or in part
abforbed by the air that is generated.
This will be well illuftrated, if we take
equal portions of water, and add to the
firft, cauftic vegetable alkali, to the fecond,

cond, aerated vegetable alkali, and to the
third an equal weight of aerated volatile alkali; upon pouring in nitrous
acid, a great degree of warmth will
arife in the firft, a moderate degree in
the fecond, and in the third one ftill lefs
confiderable; nay cold, when the quantities are varied, is produced upon fome
occafions. The reafon is, becaufe, in
the firft cafe, the acid in uniting with
the cauftic alkali gives out its fuperfluous heat without diminution; in the
fecond, the aerial acid abforbs great
part of it; and as, in the third cafe,
more of this acid is extricated, more
heat is abforbed, infomuch that the
quantity fet free being infufficient, the
deficiency is fupplied from the water
of the folution, and fenfible cold is generated. On the other hand, in the
combination of the aerial acid with
cauftic alkali, the heat neceffary to
maintain the aerial form is fet loofe,
and produces a degree of warmth correfponding

responding to its quantity. The other elastic fluids shew the same thing ; the contiguous bodies are cooled during their production, and heated when they are fixed.

3. PHLOGISTON appears *likewise to be neceſſary.* Thus the acids, as the vitriolic and the nitrous, which in a ſtate of purity contain no phlogiſton, are reſolved by warmth into elaſtic vapours, but are condenſed into drops on mere cooling ; but, by the addition of phlogiſton, afford permanently elaſtic fluids : this ſubſtance is therefore to be conſidered as a bond, affixing the neceſſary quantity of heat. The ſame thing holds with reſpect to hepatic air, which cannot be obtained from ſulphur, without the addition of phlogiſton. But the acids which always contain phlogiſton, as the muriatic, that of fluor, thoſe procured from the vegetable and animal kingdom,

kingdom, need nothing but heat to put on an aerial form.

4. MOREOVER, *the different quantity of phlogifton occafions a great change.* In nitrous air extricated from different metals, there is fome variation; that which furrounds iron filings in a clofe veffel, is by degrees fo much corrected that it not only does not extinguifh flame, but even dilates it.

IN this ftate, it is called by fome *dephlogifticated*, with what propriety I can fcarce perceive, fince the filings are at the fame time calcined, which clearly fhews a lofs of phlogifton. Befides, this experiment feems to coincide with that hypothefis which derives vital air from nitrous acid fufficiently phlogifticated. The electric fpark taken in alkaline air, produces inflammable air, either by tranfmuting part of it, or, as it feems, by fetting free the combined phlogifton,

254 On Elective Attractions.

phlogifton, and furnifhing the neceffary fpecific heat; for the bulk is much
augmented. Other inftances of variation arife from the diminution of phlogifton: fuch is that elaftic fluid which
is called *dephlogifticated fea falt*. To this
head we may alfo refer that air which
is procured from nitrated volatile alkali, digefted with magnefia nigra, and
which refembles corrupted air. The
power of magnefia nigra in dephlogifticating other fubftances is well known.
That which is generated by the explofion of fulminating gold, in which fome
degree of dephlogiftication certainly
takes place, is of the fame nature *.

ALTHOUGH the quantity of phlogifton in various bodies fhould decreafe,
as the fpecific fire increafes, I would be
far from deducing from this circum-
ftance any mutual repulfion. The aug-
mentations and diminutions compared
together do not warrant fuch a conclufion;

* Opufc. vol. ii. pp. 161, 162.

fion ; and fpirit of wine, a fubftance abounding in phlogifton, has a greater quantity of fpecific heat than water, not to mention other arguments.

BEFORE I leave the genealogy of ae-riform fluids, I muft explain what I mean by the phrafe *aerial form.* I un-derftand by this term fuch a fubtiliza-tion of a body as renders it elaftic, pel-lucid, invifible, light and permanent in cold, though not capable of paffing through the pores of glafs. *Vapours* which conftitute imperfect kinds of air, are condenfed by refrigeration. On the other hand, we have elaftic fluids, which may be not improperly ftyled *æthereal;* to fuch, neither the pores of glafs, nor of any other known body, are impenetrable. To thefe belong the matter of heat, and the magnetic fluid. The electrical fluid eafily penetrates all bodies, except electrics *per fe.* More-over, light feems to be fomething intermediate

mediate between aerial and æthereal
fubftances; for it paffes through the
pores of glafs, but not thofe of metals
and other opaque fubftances.

HENCE the neceffity of determining
the weight of the fpecific heat, in the
analyfis of aeriform fluids, plainly appears. With refpect to inflammable
air, I have before offered a fketch; and
I truft that Mr Kirwan, who has fo fuc-
cefsfully engaged in this tafk, will not
neglect this important part, fince the
analyfis will be otherwife imperfect;
and when it is once known, their nature and origin will be wonderfully illuftrated.

HERE follows a table of fpecific, as
far as they have been hitherto inveftigated. I thought it proper to difpofe
them according to the three ftates of
folidity, liquidity, and fluidity. The
fpecific heat of water is denoted by
unity.

unity. Water heated to 130 degrees melts an equal weight of fnow ; but the water thus brought to a liquid ftate is at the point of congelation. It would certainly be worth while to weigh, with the utmoft exactnefs, a piece of ice in a perfectly clofe veffel, and to repeat the operation after it was melted. The ftopple muft fit, in the moft accurate manner, left any thing fhould be loft in confequence of evaporation. This experiment has not yet, as far as I know, been performed with proper care and accuracy ; it may, however, afcertain, in fome meafure, the abfolute weight of a quantity of the principle of heat correfponding to 130 degrees.

SOLIDS.

S O L I D S.

	Spec. grav.	Spec. heut.
Aerated volatile alkali,		1,851
Swedish glafs, -	2,386	0,187
Flint glafs, - -		0,174
Agate, - -	2,648	0,195
Ice, - -		0,900
Sulphur, - -		0,183
Gold, - -	19,040	0,050
Silver, - -	10,001	0,082
Mercury, - -	13,300	0,033
Lead, - -	11,456	0,042
Copper, - -	8,784	0,114
Iron, - -	7,876	0,126
Tin, - -	7,380	0,060
Bifmuth, - -	9,861	0,043
Antimony, - -	6,107	0,063
Brafs, - -	8,356	0,116
Calcined lead, -		0,068
Calcined iron, -		0,320
Calcined tin, -		0,096
A mixture of lead and tin calcined, -		0,102

Diaphoretic

Spec. grav. Spec. heat.

Diaphoretic   antimony
    wafhed,   0,220

L   I   Q   U   I   D   S.

	Spec. grav.	Spec. heat.
Pure water, - -	1,000	1,000
Clear vitriolic acid,	1,885	0,758
Dark coloured vitriolic acid, - -	1,872	0,429
Pale nitrous acid, -		0,844
Red and fmoking,	1,355	0,576
Smoking muriatic acid,	1,122	0,680
Red wine vinegar, -		0,387
Concentrated diftilled vinegar, - -		0,103
Alkali of tartar by deliquefcence,	1,346	0,759
Cauftic volatile alkali,	0,997	0,708
Of vitriolated foffil alkali 1 part, in of water p. 2. 9.		0,728

Of

	Spec. grav.	Spec. heat.
Of nitrated vegetable alkali, p. 8. - -		0,646
Of muriated foffil alkali, p. 8. - -		0,832
Of muriated volatile alkali, p. 1. 5. -		0,798
Of depurated tartar, p. 237. 3. - -		0,765
Of vitriolated magnefia, p. 2. - -		0,844
Of vitriolated clay, p. 4. 45. - -		0,649
Of vitriolated iron, p. 2. 5.		0,734
Brown fugar diffolved,		1,086
Oil of olives, - -		0,710
—— linfeed, -		0,528
—— whale, (fpermaceti)		0,399
—— turpentine, -		0,472
Rectified fpirit of wine,	0,783	1,086
Volatile liver of fulphur,	0,818	0,994

FLUIDS.

F   L   U   I   D   S.

		Spec. grav.	*Spec. heat.*
Vital air,	-	000,132	87,000
Atmofpheric,	-	000,125	18,000
Aerial acid,	-	000,181	0,270

I HAVE been informed by Mr Kirwan, in a letter, that Dr Crawford found the fpecific heat in equal bulks of inflammable and atmofpheric air equal. Admitting this, if the fpecies of air in Mr Kirwan's table, publifhed by Mr Magellan, are eftimated by weight, the fpecific heat of inflammable air will be 281, which is more than triple of that of vital air.

D.] The theory of the diftribution being now in fome meafure explained, it is proper to confider the origin of

fire

fire in *inorganic* bodies, its propagation, and the confequences.

FIRE is fuch an accumulation of heat that the bodies expofed to it become ignited or inflamed. The chief means of exciting it are :

1. THE ftriking of flint or pyrites againft fteel, by which the abraded globules are ignited, fufed and calcined.

2. THE forging of iron, which is brought to ignition by repeated ftrokes of the hammer.

3. THE mixing of fulphur and fteel filings, which, with a proper degree of moifture, grow warm, and burft afterwards into flame.

4. ADDING the fmoking nitrous acid in a proper manner to oils.

5.

5. Pyrophorus grows red hot in atmofpheric air, and produces flame in vital air *.

In all thefe cafes, there is no heat produced without vital air, which alfo, according to circumftances, is more or lefs diminifhed. Phlogifton is likewife prefent in them all. The late Englifh philofophers contend, that the difengaged phlogifton unites with vital air, and forms aerial acid, or when in a larger proportion corrupted air, by which change, a great quantity of fpecific heat is neceffarily fet loofe, and being accumulated, produces ignition, and even flame when inflammable air is prefent. · Scheele contends, that vital air may be totally changed by phlogifton into the matter of heat. Both opinions

---

* Mr Kirwan has lately informed me by letter that there is a certain kind of earth found in Derbyfhire which takes fire in a fhort time, on the addition of linfeed oil. I have not as yet feen this earth.

opinions are fupported by ftrong arguments. It is, therefore, of great importance, that the nature of the combination of phlogifton and vital air
fhould be demonftrated. Mr Kirwan
thinks the aerial acid is the product;
and by his fagacity, has been able to
render this opinion very probable :
there is, however, ftill room for fome
doubts ; when thefe have been removed, the fyftem of Scheele will fcarce
be tenable. It is probable, that, in the
two firft cafes, a part of the fpecific
heat is expreffed by the compreffion of
the pores, and accumulated when the
dephlogifticating temperature commen-
ces, and is afterwards increafed by the
furrounding air. The ignition or inflammation of inorganic bodies is pro-
pagated to others that are capable of
it, by contact or vicinity. All bodies
may be ignited; a few only can be inflamed ; fuch are fulphur, phofphorus,
                              inflammable

inflammable air, arfenic, zinc, and fome others.

THE confequences of ignition or deflagration vary according to the diverfity of bodies, and the degrees of fire. The accumulation of heat caufes expanfion, dries, liquefies, makes bodies red hot, expels fuch as are volatile, indurates, inflames, volatilizes, calcines, vitrifies, reduces. The refidua abforb a quantity of heat fuitable to their nature, as is very evident from pyrophorus, which deftroys, during its ignition, nearly $\frac{3}{4}$ of atmofpheric air, whereas other mediums fcarce diminifh it by $\frac{1}{5}$: in vital air it takes flame, and nearly $\frac{143}{144}$ difappear *. The caufe is to be fought in the hepar which is contained in pyrophorus, and in the decompofition of the fulphur, in confequence of which the vitriolic acid is laid bare, and muft neceffarily be furnifhed with its portion

* Lavoifier, l. c.

portion of fpecific heat *. Some refidua
are of an acid nature and deliquefcent.
Others are found in the form of calces,
afhes, faline or refinous compounds.

E.] *Vegetables,* though they yield
the greateft quantity of combuftible
matter in our globe, and eafily pro-
pagate fire once lighted, generate it
themfelves but very flowly. To produce it by friction, there is required
wood, hard, dry, and penetrated with
much inflammable matter, and even, in
thefe circumftances, the operation is a
trial of patience. It is faid, that fire
has often arifen from the accumulation
of grafs not fufficiently dried, but no
one yet, as far as I know, has examined
thefe phænomena with due care.

WHEN dry wood is expofed to fire, it
grows warm and expands ; the humidi-
ty, which is generally acid, is refolved
into

* See the fubdivifion marked C,

into vapour ; the phlogiston is difpofed
to be driven off, and then is attracted,
partly by the furrounding vital air, and
is partly difengaged with the requifite
fpecific heat in the form of inflammable air, and, in the fame inftant, is
fet on fire and produces flame. Thus,
the whole fabric of the wood is gradually deftroyed, and, in the mean
time, more or lefs fmoke is emitted, by
which the foot is depofited. Fire cannot fubfift without vital air. Hence,
when it is lighted, an afflux of air in
a continual ftream takes place, which
rifes loaded with phlogifton, and rarefied by warmth, carrying along with it
carbonaceous particles not fufficiently
burned ; which particles are loaded
with fixed vegetable as well as volatile
alkali, and with earth and fal ammoniac. I have afferted, that the particles
not fufficiently burned, generate foot ;
for the furnace, which has the name of

ακάπνℨ,

ακάπνος, totally deſtroys the ſmoke which
is brought through the fire-place.

WHEN the inflammable air, and, of
courſe, the flame fail, the conflagration
is diminiſhed, and nothing but charcoal and aſhes remain. In a cloſe veſſel the charcoal amounts to one-fourth,
ſeldom to one-fifth of the weight of
the wood ; but in the open air, great
part of the charcoal is reſolved into
aſhes. This is effected by a double attraction ; the vital air ſolicits the phlogiſton, while the alkali and alkaline
earth attract the aerial acid. In a
cloſe veſſel charcoal reſiſts the moſt
intenſe fire. This ſubſtance is nothing
but a combination of phlogiſton and
aerial acid, a ſpecies of ſulphur which
is found intimately combined with
cauſtic alkali and alkaline earth, and
hence it approaches, in ſome meaſure,
to the nature of hepar. By combuſtion
in the open air, 100 parts of charcoal
contain

contain about 6 of aſhes ; of theſe $\frac{1}{900}$ conſiſts of alkali, the reſt of earth, in great meaſure alkaline. I have particularly examined well burned charcoal made of the pinus ſylveſtris of Linnæus. Of this 1 part, during detonation, alkalizes 3 of nitre, hence the phlogiſton it contains, is, in compariſon of that contained in forged iron, as 3 to $\frac{1}{2}$ *. Upon 100 parts reduced to powder, I poured concentrated vitriolic acid, then diſtilled to dryneſs, and, in a pneumatic apparatus, collected about 82 cubic inches of aerial acid, of which each is equal in weight to a docimaſtic pound. If then we add 3, which nearly correſpond to 6, the parts of the aerated aſhes, there remain 15, which give the weight of phlogiſton. No veſtiges of vital air appear. The proportion of the principles varies according to the diverſity of the wood, age, exſiccation, and

* Analyſis ferri, pp. 51, 52.

and combuftion.　Meanwhile, it appears from preceding obfervations, that
1 part of phlogifton can fix nearly
5½ of aerial acid.　The fpecific heat
of wood is a little greater than the fpecific heat of charcoal, of equal weight.

Moreover charcoal has a peculiar
attraction for elaftic fluids.　Ignited
charcoal abforbs, during refrigeration,
about 6 times its own bulk of any kind
of air.　If the hot coal be cooled in
quickfilver, or in a void fpace, it neverthelefs retains this power of abforption,
and, when immerfed in air, exerts it
inftantaneoufly, as the celebrated Fontana has obferved.　This air is expelled
by water or any other liquid.

The parts of animals are, in like
manner, confumed and converted into
charcoal, which, however, feems to a-
bound more in phlogifton, and to retain it more obftinately.　I have not
yet

yet attempted to analyfe it. Soot from animal fubftances generally abounds more in volatile alkali and fal ammoniac, than that obtained from vegetables. The fmell of the fmoke is exceedingly offenfive.

F.] In animals with hot blood, we find a fenfible temperature fuperior to that of the atmofphere, which is alfo conftant and permanent, though the furrounding medium, from its greater coolnefs, ought to abforb the difference, as happens in the cafe of other bodies. This phænomenon indicates a caufe perpetually in action, which, in this age, has been fought by many, but feems to have been moft fucceffively inveftigated by Dr Crawford. Many confiderations tend to fhew, that animal heat is generated in the lungs. The larger thefe are, the hotter are the animals. Hence birds exceed animals in this refpect. Refpiration is accelerated by

a

a more violent motion than ufual,
and the heat is at the fame time increafed, an effect to be afcribed to the
quantity of air refpired, and by no
means to friction in a body full of liquids.   Animals without lungs have
their temperature depending on that
of the medium in which they live ; to
us they feel cold, not to adduce any
more proofs of the efficacy of the air.
But the manner in which the effect is
produced, is connected with the prefent enquiry.   Dr Prieftley contends
that common air ferves to carry off
the fuperfluous phlogifton of the body.
Dr Crawford afterwards embraced this
opinion, and has admirably illuftrated
the whole procefs.   Having not yet
feen his pamphlet, I have borrowed
my idea of his fyftem from Mr Magellan's effay.   The fpecific heat of
common air is to that of aerial acid, as
69 : 1, fo that if thefe two fluids were to
receive an addition of heat of one degree

gree, the former would fix 69 times more than the latter. Now we know, from the obfervations made at Peterf-burgh, that the mercury may defcend at leaft 111 degrees of the Swedifh thermometer, below the mean temperature. Therefore, if the common air was to be changed into aerial acid, $69 \times 111 = 7659$ degrees of heat muft be fet free, *i. e.* 13 times more than is neceffary to turn iron red hot. Now, as common air is phlogifticated in the lungs, and converted into aerial acid, we may hence form fome judgment of the quantity of heat, which is fet free by refpiration, and may ferve to warm the body. Befides, as the fpecific heat of the arterial blood is to that of the venous, as 100 : 89, the author thinks it evident, that phlo-gifton is gradually accumulated in the veins, and exonerated in the lungs,

that

that the blood may be rendered capable
of receiving the due specific heat.

ALL this is pretty consistent, and
highly ingenious: but that all the fundamental parts of this theory, which
are certainly not improbable, may be
further illustrated by new experiments,
I shall add a few remarks.

1. IT remains to be ascertained, whether animals or vegetables contain most
phlogiston.   I know no experiment
which decides this question, and it
seems indeed very difficult of solution,
for the problem is, not concerning
this or that particular part, but to
compare the whole body of an animal
with a vegetable of the same weight.
If we consider our food, consisting entirely of organised bodies, we shall find
that they have been for some time dead,
before they appear on our tables, often long before; and there can be no
doubt,

doubt, but more or lefs of phlogifton
is extricated during the interval. Next
the operations of cookery diffipate a
confiderable quantity, nor can we be
certain that this is compenfated by the
fauces. The flatulency that is expelled
from the belly, is inflammable, and
the folid excretions are well known to
abound with phlogifton. Befides, a
great quantity of phlogifton feems to
be requifite for the purpofes of the
animal œconomy at all times, and in
every part of the fyftem, and I confefs
that I know not whether that which is
taken in with the ingefta is fufficient.
Still lefs can I be certain that there is
any fuperfluity to be carried off.

2. EXPERIMENT fhews a greater fpecific heat in the arterial than the venous
blood. Let us grant that the acceffion of phlogifton often leffens the fpecific fire, it by no means follows that
phlogifton

phlogifton is the agent in the prefent
cafe. We have other means of bring-
ing about the fame diminution. Thus
the pureft vitriolic acid, added to water, excites a great heat. The water
combined with the acid cannot retain
all its former fpecific heat, wherefore
the fuperfluous part is fet free, and that
without the aid of phlogifton. While
the blood is circulating, various changes
may diminifh the fpecific heat, which
indeed feems neceffary, that the parts
at a diftance from the heart, may con-
tinually receive fome heat.

3. I know no experiment which directly fhews that the blood imparts
phlogifton to the air. The air is indeed corrupted, but that this can only
be effected by phlogifton, is a mere
fuppofition. On the contrary, by
the contact of blood, nitrous air is
dephlogifticated,

dephlogifticated, and atmofpheric air
is meliorated *.

4. THAT the expired air contains a
portion of aerial acid cannot be doubted, but I think that the quantity requires to be determined more accurately. If all that is good is converted
into this acid, 1000 cubic inches of
atmofpheric air, of which one $\frac{1}{3}$ is vital,
ought, according to the analyfis of Mr
Kirwan, to be condenfed to the bulk of
about 926, *i. e.* $\frac{1}{14}$ fhould difappear, and
of vital air, 1000 fhould be reduced by
refpiration to 863, *i. e.* they fhould be
diminifhed by $\frac{1}{3}$, if they can be inhaled
by the lungs till they are totally corrupted. Suppofing that aerial acid is
further changed into corrupted air, a
fmaller contraction may be expected.

BUT the doctrine concerning the o-
rigin of animal heat, is reducible to the
fundamental

* Dr Prieftley.

fundamental queſtion, concerning the change of vital air into aerial acid, and of this into corrupted air ; an opini-on which every day ſeems to receive confirmation. Mr Kirwan has lately communicated to me a new experi-ment of great importance, made by Dr Prieſtley : That calx of mercury, which is commonly called precipitate *per ſe*, and which, when expoſed to fire, yields pure vital air, afforded but $\frac{1}{6}$ of vital air on the addition of iron filings, and $\frac{5}{6}$ of aerial acid. If the filings had ſuffered no degree of calcination, this reſult ſeems to decide the queſtion. Meanwhile, I rejoice, that it is reduced to ſuch a ſtate, that we cannot long re-main uncertain.

XLIX.

XLIX.

Column Thirty-eighth, Sulphur.

SULPHUR prefers fixed alkalis to earths; on which account, hepar made with lime, and diffolved in water, is decompofed by alkali, and a faline hepar is formed. Between vegetable and mineral alkali no difference in this refpect has yet been obferved. The power likewife of ponderous earth has not been afcertained; it probably yields to fixed alkali, for volatile alkali, both cauftic and aerated, precipitates hepar made with lime, when diffolved in a fmall quantity of water, and feparates the calcareous earth. Let magnefia alba be put into a phial with flowers of fulphur, and diftilled water; let the phial be clofely ftopped, and then digefted a few hours in a water bath; when it is afterwards cooled, it will yield a

weak

weak folution, emitting an hepatic
fmell, and turning black on the addition of nitrated filver, or acetated lead.

VOLATILE hepar, obtained from fulphur diftilled with fal ammoniac and
lime, is very foon decompofed in the
open air, fince pure volatile alkali attracts the aerial acid in preference to
fulphur. That this alkali is fuperior
to earths, appears from what has been
faid above. It is well known, that mercury and arfenic take fulphur from it,
for thefe metals, and even their calces,
when added to the volatile hepar, are
mineralized in the moift way ; the former yielding cinnabar, and the latter
red arfenic. It is probable, that this
is true of other calcined metals.
Hence it appears, that the calces of
metals may be combined with fulphur, a
truth which ochre of iron fhews clearly
and directly ; for by being mixed with
fulphur, it yields eſiorefcent vitriol : the
operation

operation may be forwarded by moiſt-
ening the mixture. It has been elſewhere ſhewn *, that calx of antimony
can take up ſulphur. Nay, the calces
of lead, tin, and ſilver, when added to
ſaline hepar, ſeem to take ſulphur from
vegetable alkali.

It now ſeems no longer doubtful,
where the oils ſhould ſtand in columns
26 and 27 ; for I have obſerved, that
a drop of oil, added to either ſaline or
earthy hepar, produces white coagula,
reſembling ſoap. This matter is alſo
ſoluble in ſpirit of wine, and the oil
may be precipitated from ſuch a ſolution by water. But it is as yet undeter-
mined in what order ſulphur attracts
the oils.

In the dry way, alkali occupies the
firſt place ; then follow the metals, of
which the reſpective ſtation is to be
ascertained

* De antim. ſulphur. p. 177.

afcertained by their mutual precipitations; but as diftinct reguli are feldom obtained by thefe means, the operation ought to be frequently repeated,
that the truth may be fully determined : the places of nickle and of cobalt
are as yet uncertain. Gold, platina,
zinc, and perhaps manganefe, refufe to
unite with fulphur, unlefs they be con-
joined with fome proper additament.

L.

Column Thirty-ninth, Saline Hepar.

SALINE liver of fulphur has no place
here, except in thofe cafes in which it
fuffers no decompofition. It diffolves
and retains almoft all the metals, zinc
alone excepted ; but no one has afcertained with what force it attracts them,
and whether they can be mutually
precipitated in the dry way. They
can

can feldom be combined without fuli-
on ; but when the combination is once
formed, it is foluble in water. Mercury, however, and antimony, are diffolved in the moift way, with the affiftance
of heat, which has not yet been afcertained with refpect to any other.

HEPAR diffolves charcoal both in the
dry and moift way ; the folution is of
a green colour.

IN the dry way, the metals precipitate one another more diftinctly than
when combined with fulphur ; the operations fhould be repeated oftener than
once, that we may be quite certain of
the conclufions ; and though I have feveral times made fuch experiments, I
have not yet attained fo much certainty as I could have wifhed. Meanwhile, I follow the order which my experiments have fuggefted ; I leave it to
be

be confirmed or corrected by future trials.

MANGANESE feems to have the fame attractive force as iron; at leaſt I have not yet been able to feparate them by means of hepar. Next follow, iron, copper, tin, lead, filver, gold, antimony, cobalt, nickle, biſmuth, mercury, and arſenic. The places of the two laſt particularly are doubtful; nor are thoſe of gold and antimony ſatisfactorily fettled.

LI.

Column Fortieth, Spirit of Wine.

I HERE ſuppoſe the ſpirit deprived, as much as poſſible, of ſuperfluous water; that I mean which does not enter into its compoſition. It attracts water very

forcibly,

forcibly, infomuch that æther diffolved in it is feparated, at leaft in great meafure. Effential oils feem to adhere to it with lefs force than æther. It takes up pure alkalis, and hepar, but the order is as yet unfettled. The Count de Lauraguais has fhewn how the vapours of fulphur may be diffolved in fpirit of wine.

LII.

Column Forty-firft, Æther.

THIS, as it were, intermediate fubftance between fpirit of wine and effential oils, forcibly attracts fpirit of wine, effential and unctuous oils. I cannot yet eftablifh with certainty the fuperior force of either of the firft mentioned fubftances. Such is its fubtilty, that it diffolves the elaftic refin, which, as

well

well as fulphur, may be precipitated by
water.

LIII.

Column Forty-fecond, Effential Oil.

THESE oils take up æther, fpirit of
wine, and fulphur, but the feries has
not been fufficiently examined ; nor
can this eafily be done, fince they do
not precipitate each other, but form
triple compounds.

LIV.

Column Forty-third, Unctuous Oil.

FIVE fubftances occur here, but their
places, if we except the laft, are not
clearly fixed.   Some acids take up em-
pyreumatic vegetable oils.   Spirit of
                                    tartar,

tartar, as it is called, abforbs no fmall
portion of oil of tartar ; and therefore,
at the conclufion of the diftillation,
they fhould be feparated, otherwife the
oil will be fenfibly diminifhed by the
acid. Vinegar has likewife this power.

LV.

Column Forty-fourth, Gold.

I HAVE already feveral times noticed
the difference between the noble and
ignoble metals. The king of metals,
to fpeak with the ancients, is dire<tly
attacked by dephlogifticated muriatic
acid, (XVII.) by aqua regia, (XVIII.)
and nitrous acid, (XIV.) ; but the other
acids, being deficient in power to carry
off the neceffary quantity of phlogifton,
do not take it up, unlefs it has been
precipitated from fome one of the three
juft mentioned. That a precipitate
procured

procured by alkali is a true calx of gold, is evident from the want of brilliancy, its folubility in aqua regia without producing red fumes, its power of tinging glafs, &c. The calx is diffolved by the acids of vitriol, arfenic, fluor, tartar, phofphorus, fat, and above all, by the acid of fea-falt in its entire ftate; but the feries remains to be afcertained. The acid of ants has not this power, at leaft it does not turn yellow; the calx, however, foon grows black, but is not reduced, fince it is taken up by muriatic acid. The fame thing is true of vinegar; but inftead of a black colour, an obfcure purple is produced. The acid of Pruffian blue, faturated with calcareous earth, precipitates gold from aqua regia, in the form of white powder; but when too much is added, it diffolves the fediment. The powder of gold precipitated by alkali, in like manner grows white, when

when put into the acid of Pruffian
blue.

ÆTHER takes gold from all the
acids. It alſo directly diffolves the
calx, leaving gold itſelf, however minutely divided, quite untouched.

CALCINED gold ſeems moreover to
be ſoluble in alkali; for when it is added to a ſolution of gold, ſo as to exceed the point of ſaturation, there ſtill
remains in the ſolution enough of the
metal to produce a diſtinct yellow colour.

IN the dry way, gold combines with
all the metals; but in what order they
are to be placed, can ſcarce be diſcovered,
ſince three and more eaſily unite without the excluſion of any one, (VIII.). I
have, however, placed thoſe uppermoſt
to which it ſeems moſt willingly, and
thoſe below to which it ſeems more re-
luctantly

luctantly to unite. The fame thing
holds with refpect to moft other metals,
concerning which let this admonition
fuffice.

GOLD is foluble in faline hepar.
though it rejects fulphur.

LVI.

Column Forty-fifth, Platina.

WHAT has been juft faid of gold, is
applicable in great meafure to platina,
which, however, in the ftate of a precipitate, is foluble in more acids, as in
that of fugar, forrel, lemon, ants, and
in vinegar. The acid of Pruffian blue
feems to have no power either as a precipitant or a folvent.

THAT platina is always contaminated with iron, in my opinion, indicates
nothing

nothing but the prefence of both metals in the places where platina is found. He alfo who fhall confider the great difficulty with which platina is fufed, will not wonder that the alloy is defended by it fo as fcarce to be feparable. This is ftrongly confirmed by the precipitate of platina from aqua regia by fal ammoniac, which fhews no veftiges of iron, when it is well fufed in microcofmic falt *. It feems moft probable that the magnetic power of the inherent iron is acquired by the triture in the iron mould, while the gold is amalgamated; it is at leaft by this means contaminated with quickfilver. Scarce any platina is brought to Europe, which has not firft undergone this operation. Gold mixed with iron in fuch a proportion as to equal platina in fpecific gravity, totally differs from it.

THE

* Opufc. vol. ii. p. 179.—181.

THE experiments of the celebrated Dr Lewis feem to indicate that platina is in fome degree attacked by liver of fulphur.

LVII.

Column Forty-fixth, Silver.

MURIATIC acid attracts filver more ftrongly than any other, and takes it from all the reft.   The acid of fat, however, feems to equal it.   It is probable that the acid of Pruffian blue is fuperior to none but the aerial.   By the former faturated with lime, filver is precipitated from vitriolic and nitrous acid, in the form of a white powder, but is rediffolved when too much is added.   The acid of fugar feems to come next that of fat, for it decompofes lunar vitriol by attracting its metallic bafis : nitrated filver is precipitated by the

vitriolic

vitriolic acid, by that of fugar of milk, and likewife by the arfenical acid, but fo imperfectly that it fhould, in appearance, be placed after that of nitre. The places of the following acids are lefs certainly determined. Silver precipitated by cryftallized alkali is foluble in aerial acid, which may again be expelled by fire ; aerated filver, however, is not taken up by water. Vitriolated filver is not precipitated by aerial acid, unlefs it contain a mixture of muriatic acid.

PURE volatile alkali diffolves calcined filver, and the folution will afford cryftals. There is a new clafs of falts, confifting of metals diffolved in alkalis, highly worthy of attention, though they have as yet been but little, or not at all, examined.

LVIII.

LVIII.

Column Forty-feventh, Mercury.

MERCURY, in point of fufibility, conftitutes one extreme among the metals,
and platina the other. The former requires only fuch a degree of heat as is
rarely wanting in our atmofphere, but
when the cold is increafed by art to
the temperature denoted by 40° of the
Swedifh thermometer, this metal likewife begins to concrete, and, in due
time, becomes quite hard. Dr Pallas
fays that it was feveral times congeal-
ed in Siberia by the natural cold. In
its common ftate, therefore, it is to be
confidered as a metal in fufion; and
fince, in its folid ftate, it is nearly as
malleable as lead, it by no means ought
to be placed among the femimetals,
otherwife the whole clafs muft be confidered

fidered as brittle, for none is malleable
when in fufion.

Acid of fat is placed firft, for it difengages all the reft, even the muriatic,
to which the fecond place belongs.
The acids of fugar, forrel, amber, arfenic, and phofphorus, foon expel the vitriolic and nitrous, and fall with the
calx of quickfilver to the bottom ; but
their refpective forces have not been
fufficiently compared : acid of fugar of
milk precipitates mercury, but yields
to the muriatic, but whether to the vitriolic and thofe yet ftronger, is uncertain. Acid of lemon produces a copious precipitation of mercury, diffolved
in the cold in nitrous acid, though but
a fparing one when the folution is forwarded by heat. The fame holds with
refpect to the acid of tartar, of which
it is moreover certain that it yields to
the vitriolic. The fluor acid feems to
be weaker than the nitrous : the acid
of

of ants does not, as we learn from Margraaf, diffolve, but reduce the calx.
The ftations of acetous acid, phlogifticated vitriolic acid, and the acid of borax, remain to be afcertained with
greater accuracy. The calx of mercury, precipitated by mild alkali, combines with the aerial acid; but this metallic falt is not foluble in water. The
acid of Pruffian blue decompofes aerated mercury, and forms cryftals. This
acid precipitates filver from its folution
in nitrous acid, when made in the cold,
in the form of a black powder. Whether it prevails over the vitriolic, and
thofe ftill ftronger, by its fingle power,
has not yet been determined by experiment.

LIX.

LIX.

Column Forty-eighth, Lead.

THE vitriolic acid attracts lead with greater force than any other, and immediately takes it from them. The acid of Pruffian blue, alone, has no power ; but by a double attraction a white powder is feparated, which cannot be rediffolved by adding an excefs of the precipitant. The acids of fat, of fugar of milk, fugar, arfenic, tartar, phofphorus, and forrel, certainly expel the muriatic and nitrous acids, at the fame time forming new compounds fcarce foluble ; but their refpective order requires to be afcertained by farther examination. The fluor acid prevails over vinegar, as alfo do probably the acids of lemon and ants. To the reft, the obfervations in LVIII. are applicable. The calx of lead, when it contains

tains no aerial acid, feems to attract it
with the fame force as fixed alkali, for
the calx in this ftate renders the alkali
cauftic in part, as reciprocally happens
to the aerated calx when put into the
cauftic ley.

PURE fixed alkali, and alfo unctuous oil, diffolve the calx of lead.

LX.

Column Forty-ninth, Copper.

ACID of fugar occupies the higheft
rectangle, fince when it is dropped into vitriolated or muriated copper, it
feizes the metal, and exhibits at the
bottom of the veffel a greenifh fky-blue
powder. Acid of tartar likewife precipitates thefe falts, but not fo quickly;
it forms blue cryftals. The muriatic
acid is fuperior to the vitriolic, for blue
vitriol

vitriol is readily diffolved in it; the
menftruum foon grows green, and yields
a yellow fympathetic ink, which cannot
be obtained without muriated copper;
not however to truft to the colour alone,
I add highly rectified fpirit of wine
to muriatic acid, faturated with vitriol;
no feparation, however, took place, as
neceffarily happens whenever copper is
united with vitriolic acid. At the fame
time, let it be obferved, that a very
fmall degree of heat, even the rays of
the fun, reftore the fuperiority to vitriolic acid, infomuch, that cryftals of vitriol are at laft obtained, or may indeed
be feparated by fpirit of wine without
evaporation. This is a remarkable inftance of the power of heat, (IV.). The
muriatic acid dropped into a folution
of nitrated copper, precipitates a white
faline powder, confifting of marine a-
cid and the calx of copper in excefs:
this powder is not foluble in water.
The acid of fat, of fugar of milk, and

of

of nitre, are expelled by the vitriolic,
and the acetous by the arfenical; but
the ftrength of the reft has not been
fufficiently examined. The Pruffian a-
cid, without affiftance, decompofes aerated copper; but fcarce any other
compound of this metal. When combined with alkali, it decompofes them
all, by means of a double attraction,
and the precipitates are rediffolved
when too much is added. The other
acids take up only part of thefe fediments; what remains is of a white
colour. Volatile alkali totally diffolves
them; the colour of the folution is a
bluifh green, but they are again precipitated by water.

ALKALIS and oils attack copper,
but in what order is not known.

LXI.

LXI.

Column Fiftieth, Iron.

ACID of fugar immediately turns a folution of martial vitriol yellow, and gradually feparates a yellow powder, confifting of the calx of iron, and the added acid. The acid of tartar, in like manner, decompofes it ; but the new falt does not fo foon become vifible, and it is more cryftalline. Green vitriol, diffolved in muriatic acid, is feparated by fpirit of wine, and therefore the vitriolic is to be placed firft. Acid of fugar of milk is incapable of feparating the vitriolic acid ; and the acid of fat yields to the nitrous. Pruffian blue is diffolved by its own acid ; the folution is of a yellow colour : other acids have no action upon it, which would feem to fhew that this acid has

the

the ſtrongeſt attraction for iron ; and it
does indeed precipitate it from the aerial
acid, but from no other, as far as I know,
unleſs it is ſaturated with alkali, that
is, by means of a double elective attraction. The following places remain
to be confirmed by farther experiment ;
it is, however, certain that the acetous
acid is inferior to the arſenical.

LXII.

Column Fifty-firſt, Tin.

IN almoſt the whole of this column
the ſeries is doubtful, and very difficult to be aſcertained, ſince tin requires an exceſs of acid to be ſuſpend-
ed. It is certain that the acids of arſenic, and ſugar of milk, yield to the vitriolic and marine, while they are ſu-
perior to the acetous. The acid of
fat

fat exceeds the marine in ſtrength of attraction.

BOTH fixed and volatile alkali attack calx of tin.

LXIII.

Column Fifty-ſecond, Biſmuth.

BISMUTH readily diſſolves in nitrous acid; but the acids of ſugar, fat, ſorrel, tartar, phoſphorus, and arſenic, when added to this compound, attract the baſis; but their relative powers are un-determined: the new compounds fall to the bottom ſcarce ſoluble, in the form of very fine powder, except only tartarized biſmuth, which, however, affords pellucid cryſtalline grains, in 10—15 minutes. As water alone cauſes a precipitation, I either employed a ſo-lution with ſuch an exceſs of acid, that

a

a number of drops of water, equal to that
of the precipitants, caufed no permanent
cloudinefs, or elfe added acids, which
may be procured in a concrete form,
as moft of thofe juft mentioned. Thefe
acids, in like manner, decompofe a folution of bifmuth in the vitriolic acid :
which menftruum, when diluted, attacks
the calx ; but to diffolve the regulus,
it muft be in a concentrated ftate ; and,
in order to feparate the phlogifton, it
muft be evaporated to drynefs.

Distilled vinegar boiled with the
calx of bifmuth for half an hour, does,
in reality, diffolve part, as appears from
the tafte, the addition of phlogifticated
alkali, and the above mentioned acids ;
what is diffolved cannot be precipitated by water, unlefs perhaps in great
quantity, and by long ftanding. The
regulus is diffolved in the fame manner, but fo fparingly that it can fcarce
be afcertained. What has been faid
concerning

concerning vinegar, is likewife applicable to acid of ants. The remaining
places are uncertain, nor are even the
refpective powers of vitriolic, nitrous,
and marine acid determined.

LXIV.

Column Fifty-third, Nickle.

Nickle is not yet univerfally acknow-
ledged as a diftinct metal; but as it
may be diftinguifhed from others by
conftant criterions, fuch as its deep
green colour, when diffolved in thofe
acids which attack it; its blue colour
in volatile alkali; the greenifh white
precipitate it yields on the addition,
either of common or phlogifticated
alkali; the hyacinthine tinge it communicates to glafs, characterifics which,
taken together, belong to no other;
moreover, fince when it is properly
purified,

purified, it cannot be refolved into others, though it be ever fo long tortured, both in the dry and moift way; laftly, fince no one has produced by fynthefis, a mixture agreeing with nickle in the properties above mentioned, from copper, arfenic, purified cobalt, iron, or other metals fufed together; for thefe feveral reafons, I cannot but confider nickle as a diftinct metal, till I am better informed by new experiments. Moft chemifts have been feduced by the extreme difficulty which attends the purification of it. It is indeed always contaminated with arfenic, cobalt and iron, fometimes alfo with copper and other metals. Copper is eafily feparated, arfenic with great difficulty, the laft veftiges of cobalt with ftill greater, but iron by no method hitherto difcovered, as is related more at length in my differtation on this metal. I do not, therefore, wonder

wonder that nickle, if it be fo fparingly contaminated with cobalt, that the
particles of the former metal furround
thofe of the latter on every fide, fhould
not afford, according to the common
method, glafs of a green colour, and
yet that this colour fhould appear on
the addition of white arfenic ; for this
addition not only weakens the cohefion
of the cobalt and nickle, but renders
the mafs more fluid, and deprives the
cobalt of the phlogifton which before
prevented the effect. Cobalt does not
tinge glafs, except when in the ftate of
calx ; this calx contains a wonderful
ftore of colours ; but when exceffively
dephlogifticated, it cannot either be
fufed or reduced, without great difficulty. Nickle contaminated with iron
alone, which I have not been able to
remove by any means, is malleable, and
very tenacious, fo that I doubt whether it
ought to be reckoned among the brittle
metals. It is fometimes magnetic, is
difficult

difficult of fufion, and docs not yield a blue glafs on the addition of white arfenic ; it, however, gives a very deep green to acid menftrua, and fhews the above mentioned criterions.

NICKLE prefers no acid to that of fugar ; by this it is taken from every other, and appears in the form of a whitifh green infoluble powder : It is likewife precipitated by acid of forrel. The acid of fat yields to the nitrous. The other places remain to be determined by farther examination ; the trials, however, that have been made, feem to indicate that the arfenical acid is to be placed after the acetous.

LXV.

LXV.

Column Fifty-fourth, Arfenic.

THE folutions of arfenic are, in fome meafure, imperfect, at which we need not be furprifed, the calx being only a real acid, coagulated by phlogifton, (XX.). It has, however, as yet been little examined, with a view to its elective attractions. That the vitriolic acid yields to the muriatic, appears from this, that arfenic, diffolved in the former, yields, upon addition of the latter, and expofure to a very gentle heat, butter of arfenic. The vitriolic is likewife extruded by the faccharine, and the febaceous by the nitrous. The reft is doubtful.

LXVI.

LXVI.

Column Fifty-fifth, Cobalt.

COBALT differs from nickle, in im-
parting a red colour to all the acids,
and volatile alkali, when it is diffolved
in them; in the reddifh afh-coloured
precipitate thrown down either by
common or phlogifticated alkali; in
attracting faline hepar from nickle in
the dry way; in refufing to combine
with filver, bifmuth and lead by fufion,
which metals do not reject nickle unlefs it contain too much cobalt; in the
fuperior richnefs of its colour; for
which reafon, though it be prefent in
the fame mafs, in far lefs quantity than
nickle, yet it prevails; for a regulus
containing a much larger portion of
nickle, yields neverthelefs a red folution in acids, without any fhade of
green, and with a ftill more inconfiderable

derable alloy, it tinges glafs of a blue colour.

Cobalt is moft ftrongly attracted by acid of fugar, which precipitates it from other acids, in the form of a pale rofe-coloured powder ; and as it is very difficult of folution in water, unlefs a great excefs of acid be prefent, its power of attraction has not yet been compared with that of the acid of forrel, which likewife precipitates cobalt from the muriatic, and other acids. The vitriolic is expelled by the muriatic acid, as may be fhewn in various ways. Highly rectified fpirit of wine refufes vitriol of cobalt, but not muriated cobalt. Since, therefore, a folution of this vitriol in marine acid, affords no precipitate on the addition of fpirit of wine, it is evident that the vitriol muft have been decompofed. Befides, muriated cobalt (but not vitriolated) yields fympathetic ink ;

now

now a folution of vitriol, upon the addition of muriatic acid, (or of fea-falt,
which contains it, and then the decompofition is effected by a double attraction), immediately acquires this property, and in a dry ftate of the air,
writing is turned green, and becomes
legible. I fay, when the air is dry, for
when the letters are invifible, if the
paper be put over newly burned lime,
or concentrated vitriolic acid, in a
clofe phial, they foon become manifeft.
Fire, therefore, or heat, acts only by
drying, which is agreeable to Hellot's
explanation. Cobalt precipitated with
phlogifticated alkali, is neither foluble
in phlogifticated alkali, nor acids.

Acid of arfenic is incapable of taking cobalt from vinegar, at leaft it
caufes no precipitation. The other
places remain to be further examined.

LXVII.

LXVII.

Column Fifty-sixth, Zinc.

Acid of fugar takes zinc from e-
very other acid, and when united with
it, immediately falls to the bottom in
the form of a white powder ; but the
acid of fugar of milk yields to the
vitriolic, and that of fat to the nitrous.
Zinc precipitated by phlogifticated alkali is not acted upon by an excefs of
it ; but it is taken up by acids. Vitriolic, nitrous, and muriatic acids,
prevail over that of arfenic ; but the
acetous yields to it. Vitriolic acid
comes before the muriatic ; for vitriol of zinc, diffolved in acid of falt,
is precipitated by fpirit of wine. Acid
of forrel has not been tried ; but if I
miftake not, it will be found to expel
the vitriolic.

LXVIII.

LXVIII.

Column Fifty-seventh, Antimony.

THE attractions of antimony have as yet been but little examined; the examination is indeed attended with difficulty, fince the folutions require an excefs of acid. The firft place belongs to the acid of fat, and the next to the muriatic; the vitriolic is expelled by the faccharine. The vitriolic, nitrous, and muriatic, are fuperior to the arfenical; to which, however, the acetous yields. I have not yet been able to afcertain a greater number with accuracy.

LXIX.

LXIX.

Column Fifty-eighth, Manganese.

THE fpecific gravity of magnefia nigra, the property it poffeffes of tinging glafs, and, above all, the white precipitate produced by phlogifticated alkali, in folutions made in every acid, led me, many years ago, to perceive diftinctly that this fubftance contained a peculiar metal. Dr Gahn, who was formerly my pupil, firft eliquated the regulus, which has moft diftinguifhing properties; and fince neither analytic experiments have refolved it into others, nor fynthetic compofed one with the fame properties, it is proper to diftinguifh it. Manganefe in its metallic ftate is hard, brittle, has a granular, white and fhining fracture; and fuch is its refractorinefs, that it is more difficult of fufion than iron,

whence

whence I at firſt conjectured, that it was the ſame as platina ; it ſeems to refuſe ſulphur ; it yields a perfectly pel-lucid and colourleſs vitriol, of which the cryſtals are parallelopipeds. The calx, when deprived of almoſt all its phlogi-ſton, is black ; but when it has a ſuffi-cient quantity to be capable of ſolution in acids, it is white ; when in combi-nation with a ſtill larger portion, it ac-quires a reguline nature. The black calx, in the fire, gives an hyacinthine tinge to borax, and a purple one to mi-crocoſmic ſalt ; but on the addition of a ſufficient quantity of phlogiſton, both colours diſappear. This metal parts with great difficulty from all its iron ; but who knows not the difficulty of ſe-parating the laſt veſtiges of foreign matter, when it is ſurrounded by o-ther particles, which attract them ſtrongly, eſpecially if the mixture be refractory ?

THE black calx is taken up indeed by the vitriolic and marine acids, but the folutions are coloured, and never without a tinge, unlefs an addition of fugar, or fome other matter, be added to fupply the neceffary phlogifton; but it is perfectly diffolved in acids, either artificially phlogifticated, as thofe of vitriol and nitre, or thofe naturally containing inflammable matter, as thofe of lemon and tartar; and it decompofes them at the fame time.

THE acids of fugar, tartar, forrel, lemon, phofphorus, and fluor, expel the nitrous, vitriolic, and marine; for when vitriol of manganefe is diffolved in them, there appear fmaller cryftals, eafily foluble in fpirit of wine, which totally rejects vitriol; and, moreover, the folution in which the cryftalline grains are immerfed, when poured off, afforded no precipitation on the addition of fpirit of wine. The acids of
nitre,

nitre, fat, and arfenic, expel the acetous. The reft is doubtful.

IN the dry way, copper, iron, gold,
filver, tin, and fiderite, combine with
manganefe. The other metals remain
to be tried. Liver of fulphur fcarce
feparates the alloy of iron, but diffolves
both metals together.

LXX.

Column Fifty-ninth, Siderite.

THIS metal, which renders iron cold-
fhort, feems to me to be different from
all others. The few circumftances
which I have hitherto been able to obferve concerning it, may be feen in my
effay on that fubject. Much remains
for inveftigation ; and I have been obliged to put off my refearches for want
of materials to work upon. The three
common

common mineral acids diffolve it, but
with difficulty. In the feries of precipitations by metals, fiderite feems to
ftand higher than lead. It cannot, any
more than tin, be precipitated in a metallic form, but always falls down in
the ftate of a calx.

Such is this extenfive fubject, and
fuch a multitude of experiments and obfervations does it ftill demand. I have
diftinguifhed what is certain from that
which remains doubtful, that it may appear what remains to be done by him
who wifhes to try his powers and patience in the cultivation of this fcience.
The ftations which are ambiguous or
doubtful, have not been affigned totally
without reafon, though indeed infufficient to produce full conviction. Moreover, if I have any where erred, the
condition of humanity muft plead my
excufe. I do not, however, doubt but
that

that many affertions, which fhall feem obfcure, or perhaps falfe, to fome, will be quite plain and evident to him who fhall ferioufly apply to this tafk.

E N D.

N O T E S

PRECEDING DISSERTATION.

P. 4.] CONCERNING thefe admirable experiments of M. de Morveau, we have a very acute and pertinent obfervation by an author, frequently fuperficial. He remarks, that there is a fource of error in them of which M. de Morveau was not aware, for the inferior furface of the highly polifhed plates, which are brought into contact with the mercury, being more or lefs readily diffolved by it, according to the nature of the metal, will acquire an unequal addition of matter; and hence the difference in the weights, neceffary to feparate the laminæ from the furface of the mercury, may arife, not from any difference of attractive power, but from inequality of mafs. (Fourcroy, Leçons Elem. Diff. fur les Affinit.).

P. 32.] THE fentence beginning in l. 11. would be clearer, if placed in the following manner :
" In the firft place, we remark that a portion of
" phlogifton flies off in the inflammable air," &c.

P. 47.

P. 47. l. 1. &c.] M. QUATREMERE D'ISJON-
VAL affirms, (*Collection de Memoires*, Paris 1784,
p. 219. &c.) that when folutions of muriated magnefia
and muriated lime, and likewife of vitriolated volatile
alkali and vitriolated magnefia, are mixed together,
the precipitation which takes place, is without decompofition, and the effect of the ftrong attraction of one
of the compounds for water. But the excessive ignorance this author betrays of the moft common obfervations in chemiftry, (for he affirms, that when nitrated lime and vitriolated tartar are mixed together,
the latter falt is precipitated) fhews that his opinion
is not worthy of the flighteft attention.

IT is more furprifing that Mr Morveau, the tranflator and correfpondent of Bergman, and, unqueftion-
ably, one of the moft philofophic chemifts in France,
fhould put the following queftion, five years after it
had been folved by our author, (N. Act. Upfal, V III.
1775.) " How does it happen that two falts, which
" when feparate have a fufficient quantity of water
" for their folution, fhould, upon mixture, immediate-
" ly yield cryftals, as if the water had been attracted
" by fpirit of wine? 1his a *totally new* queftion,"
&c. (l. c. p. 221.).

P. 65. l. 20.] THE words *of the fubftances* may be
ftruck out without injuring the fenfe ; and the omif-
fion would make the fentence run fmoother.

P. 77.]

P. 77.] NEITHER do the phænomena which attend the combuftion of fulphur, nor others of the fame nature, admit of rational explanation upon any principles hitherto known, unlefs we adopt Mr Cavendifh's difcovery concerning the conftituent parts of water. The experiments of that excellent chemift, in my opinion, lead to more fpeculations, and promife the folution of more phænomena, than any which have been publifhed fince the fundamental difcovery of Dr Black. But I fhall have occafion to confider them more particularly below.

P. 89. 32.] THIS conjecture of the author's would feem to be erroneous, for Mr Wiegleb (Crell's Neueft Entdeck. Th. 11. p. 14. 1783.) relates fome experiments, from which it evidently appears, that fixed vegetable alkali has a ftronger attraction for vitriolic acid, *viâ ficcâ*, than the heavy earth. ℥fs of the heavy fpar being expofed to a ftrong heat in a crucible, with ʒvi of falt of tartar, was decompofed all but 28 grains. Mr Wiegleb adds, that this is a much eafier procefs than that of Scheele and Bergman Dr Withering has fome experiments (Phil. Tr. v. lxxiv. p. 304. 1784.) that exactly coincide with thefe, but Mr Wiegleb's feem to be of an earlier date. No mention is made by either of foffil alkali; but it may be fuppofed to agree with the vegetable. I have arranged thefe fubftances accordingly, but have drawn no line between them, fince it is not abfolutely certain,

tain (though I have little doubt of it) that pure fixed alkali will effect this change. That the aerial acid comes into action in the above mentioned experiment, there can be no doubt, for Mr Wiegleb makes exprefs mention of an effervefcence.

P. 96. l. 2.] WHENEVER the author difpofes fub-ftances by conjecture or analogy, he takes care to in-form his reader. As he therefore fpeaks pofitively in the prefent paragraph, it is to be concluded that he fpeaks from experiment. Dr Withering, how-ever, (l. c.) affirms the contrary with great confi-dence. I have fo often, fays he, repeated thefe ex-periments, to fatisfy myfelf and others, that I am perfuaded the terra ponderofa cauftica ought to be placed below the alkalis, exceptin the column ap-propriated to the vitriolic acid. Mr Kirwan, confiding in the accuracy of Bergman, afks, Whether a decep-tion may not have arifen from the abforption of an excefs of acid, by the alkalis that were added ? It is likewife to be remembered, that when Dr Withering employed pure vegetable alkali, he obtained a preci-pitate, foluble neither in water nor acids, *viz.* a com-bination of the alkali and earth. The fame precipi-tate likewife appeared when an aqueous folution of pure terra ponderofa was added to pure vegetable or foffil alkali, but none when it was added to pure vo-latile alkali. Thefe precipitates are undoubtedly well worthy of farther examination. It muft furely, *a*

priori,

priori, feem juft as extraordinary, that volatile alkali
fhould throw down the heavy earth, as that the fixed alkalis fhould be precipitated by it. Meanwhile,
till this matter is thoroughly cleared up, I have placed
a note of interrogation after the heavy earth, in the
table of words.

Ibid. note.] Mr Cavendish found that the water
proceeding from the deflagration of inflammable and
dephlogifticated air, is always impregnated with nitrous
acid, whenever thofe airs are exploded in fuch a proportion, that the burnt air is not much phlogifticated, from whatever fubftance the dephlogifticated
air may have been procured. But if the proportion
be fuch, that the burnt air is almoft entirely phlogifticated, the condenfed liquor is not at all acid, but
feems pure water, without any addition whatever ;
and when they are mixed in this proportion, very little air remains, almoft all being condenfed. Thefe
phænomena may be explained, by fuppofing, either
that nitrous acid, in fmall proportion, is a conftituent
part of dephlogifticated air, or that phlogifticated air
is nitrous acid, with a larger proportion of phlogifton
than nitrous air. To the latter fuppofition Mr Cavendifh evidently inclines, and obferves, that, in con-
formity to it, part of the phlogifticated air, with
which the vital is debafed, is, in his experiment, converted into nitrous acid, by the ftrong affinity of the
latter to phlogifton. As a confirmation of his fuppofition,

pofition, he remarks, that when nitre is deflagrated with charcoal, the acid is almoft entirely converted into this kind of air.

THIS acute conjecture points to the origin of ni-trous acid, a difcovery which, fince chemifts have been fo converfant with elaftic fluids, has always feemed to be near at hand, though it has conftantly eluded their grafp. For if it be true, it is reafonable to imagine that Nature has fome procefs by which fhe difengages the acid, and perhaps, in the variety of her operations, another by which fhe again combines it. At all events, the fubject is worth profecuting. And it would feem advifable to expofe nitrous air to various fubftances, by which we may expect to com-municate phlogifton to it ; for although this has been already done, as by expofing it to liver of fulphur, and the refult has been fuch as feems rather to fa-vour Mr Cavendifh's hypothefis, yet we are not enough acquainted with it to draw a certain con-clufion.

ANOTHER method might be, to obferve the effect of vital upon phlogifticated air, under as many dif-ferent circumftances as can be imagined. Some ex-periments on inflammable air, to be mentioned here-after, would feem to afford encouragement for fuch an inveftigation. Thefe elaftic fluids are indeed con-ftantly prefent together in the atmofphere, but that is a fituation not calculated for fuch obfervations. Might not the electric fluid be of great fervice here alfo ?

WE

WE are besides indebted to Mr Becker of Magdeburg, for some recent observations on the origin of the nitrous acid. In a pamphlet, (Entdecktes Saltpeter-sauer in den Animalischen Ausleerungen, Dessau. 1783), he rejects both the ancient and modern opinions concerning the generation of this acid, either as palpably false, or as unsupported by any adequate proof. He asserts, that the putrid fermentation is not at all necessary to its production. He found (Experiment I.) nitrous acid in cows urine, which had been exposed for eight days to the sun. He mixed some of the soakings of a dunghill with a ley of burnt sheeps dung, and chalk in powder. The mixture began to ferment on the following day, and on the fourth, the internal commotion having ceased, he found at the bottom of the phial, regular crystals of prismatic nitre

IN a supplement to this publication, he tells us that he has found the full solution of the problem concerning the generation of nitre, and that the acid is not to be sought in the air, but in the vegetable kingdom, *by means of the excretions of animals.* " I found further, says he, that this kingdom affords not only the common fixed alkaline salt, but also a fixed-alkaline-animal neutral salt, which appeared on lixiviation, notwithstanding the dung was dried and burnt. It is truly surprising, that during the burning of the straw or dung, its alkali, together with the acid contained in the dung, should not be destroyed by the process, procefs, but fhould combine with each other. The farther I proceeded, the more I difcovered. On examining the earth of ftables and cow-houfes, I found that its lixivium yielded prifmatic nitre, while that of the dung would only afford fmall cryftals, which required an addition of nitre, in order to be reduced to a prifmatic form. Moreover, I can extraɗ nitre at pleafure, in the courfe of three days, from the earth of ftables and cow-houfes, by ufing for faturation well purified potafhes."

THESE experiments do not, indeed, fhew the conftituent parts of nitrous acid, but they may ferve to warn us againft falfe theories. I am forry that I can give no account of the experiments and opinions of Mr Thouvenel, the fuccefsful candidate for the prize offered by the French Academy, having never yet been able to obtain a perufal of his differtation.

P. 111. l. 15.] IT may be worth while to examine into this matter a little more narrowly. The difference between neutral falts containing phlogifticated and dephlogifticated acids, is very ftriking, in many inftances. Should we even admit, that the alkali contains a portion of the inflammable principle, and communicates it to the acid, there muft ftill be a deficiency, i. e. lefs phlogifton than in common falt, unlefs it be fupplied from fome other quarter ; and we might expeɗ a fenfible difference. It is not to be expeɗed, from what is faid of the attraɗion of vital air for

phlogifton,

phlogiſton, that the aerial acid (ſuppoſing it to con-
ſiſt of theſe two ſubſtances) of the alkali will be de-
compoſed by the dephlogiſticated muriatic acid. I
wiſh, however, that the experiment were made, with
a view to the examination of the elaſtic fluid.

SINCE this part of the note was written, I have
ſeen a paper on the dephlogiſticated marine acid, by Mr
Bertholet, (Journ. de Phyſique, Mai 1785.) who directed his attention, in ſome meaſure, to this very object. He boiled in an air-apparatus a mixture of foſſil alkali and dephlogiſticated marine acid, and found
that the diſengaged elaſtic fluid was at firſt aerial acid
with common air; next, air of a purer kind; and,
laſt of all, aerial acid again. From calcareous earth,
no aerial acid is diſengaged, but only atmoſpheric air,
which gradually becomes more and more pure, and is at
laſt very pure dephlogiſticated air. This laſt experiment
looks very like a confirmation of my conjecture, that
the dephlogiſticated acid gets phlogiſton from the elaſtic fluid. But Mr Bertholet has, by no means, ſufficiently inveſtigated the problem, though what he obſerved may ſerve ſtill further to ſhew that it is worthy
of inveſtigation.

THE atmoſpheric air, in the firſt experiment, was
beforehand contained in the veſſels. Whence the vital air that appears afterwards proceeds, it is not eaſy
to tell. Can it come from the decompoſition of water,
which perhaps the ſtrong attraction of the dephlogiſticated acid may aſſiſt in accompliſhing in a gentle heat?

Mr

Mr Bertholet himfelf, conformably with the new French hypothefis, deduces it from the dephlogifticated muriatic acid.

THE neutral falt, formed in this experiment, was exactly like common falt.

WHAT he fays of volatile alkali, is very obfcure. He perceived an effervefcence, even when the alkali was cauftic ; and the elaftic fluid was of a peculiar kind, and, as he thinks, is formed by the combination of volatile alkali, and the dephlogifticated air yielded, according to his hypothefis, by the acid.

POSSIBLY the acid, by attracting the phlogifton of the volatile alkali, may decompofe part of it ; and if fo, the elaftic fluid that is extricated will be the fame as that which is obtained by the explofion of fulminating gold, (Scheele on air and fire, Bergman Opufc. vol. ii.); and the acid being thus reduced to common marine acid, will unite with the reft of the volatile alkali and form fal ammoniac, which was the product obtained by Mr Bertholet.

P. 115. l. 11. 12. &c.] MR TILLET, who has lately (Mem. Paris. *année* 1780.) examined the action of nitrous acid upon gold, in the circumftances defcribed by Mr Brandt, has been led to form an oppofite opinion. He allows, that nitrous acid, under thefe circumftances, does actually *attack* gold *in leaves, and in a ftate of ductility*, but contends, that it does not really diffolve it either wholly or in part, keeping it only mechanically

mechanically fufpended. Whether he has brought any new experiments or arguments that prove his af-fertion fatisfactorily, let the reader judge. He ob-ferves, and it was known before, 1. *That, if a little fil-ver be added to nitrous acid containing gold, the latter me-tal will be precipitated.* The connection between me-chanical fufpenfion and this effect, is not very obvious; but, if we fuppofe the gold to be diffolved, then it may be faid, that the phlogifton afforded by the filver is the caufe of the precipitation ; fo that this phæno-menon would appear to be rather unfavourable to the author's opinion, and fo far unfavourable as to coun-terbalance all his other arguments.

2. GOLD *thus precipitated,* (1.) *notwithftanding its ten-der and fpongy ftate, is not taken up by nitrous acid, how-ever concentrated and affifted by heat.* It certainly feems extraordinary, that metallic particles, diffufed through the fubftance of another metal, fhould be, in fome mea-fure, foluble ; and yet, that thefe very particles, in a ftate of equal tenuity, fhould become infoluble, when the other metal has been removed. But the fact, how-ever remarkable, can fcarce be thought conclufive. It is the oppofite of that cafe, in which the particles of a body, eafily foluble by themfelves, are yet pre-vented from being diffolved, by being mixed with a large proportion of an infoluble body.

3. MR TILLET *found, that all the gold was depofited while the nitrous acid was paffing through a filter of four folds of paper.* Mr Brandt obferved, that the gold

was

was depofited after the acid had ftood fome time, and alfo on agitation.

4. On *examining a drop of the acid with the micro-fcope, Mr Tillet faw the particles of gold in their metallic ftate, floating in it.* Can we fuppofe, that fome parti-cles are fufpended, while others are diffolved? Or, may it be conjectured, that, as the noble calces eafily recover their phlogifton, a fource of error might arife from the expofure of the folution to the fun's rays? That fuch an accident might happen, appears from Mr Tillet's total filence with refpect to this circum-ftance.

The Commiffioners, moreover, (l. c. p. 615.) ob-ferve, that it appears from feveral of their experiments, that the pureft nitrous acid takes up (*fe charge avec*) a few particles of gold.

P. 120. l. 7. 8.] This opinion concerning the caufe of the corrofive nature of certain metallic falts, has been adopted and confirmed by many experiments by Mr Bertholet, [Journal de Medecine, 1780, p. 50. The fame effay was likewife fince reprinted, with ad-ditions, in the Mem. Par. for 1780.] Among his ex-periments, the following feem the moft conclufive : Corrofive fublimate, expofed to heat, (not a violent degree), with oil, is, for the moft part, reduced. A piece of flefh being put into a folution of this mercurial falt, a copious precipitation took place; the liquor now

reddened

reddened fyrup of violets, whereas it had before turned it green. The precipitate was calomel.

PRECIPITATES of corrofive fublimate, whether with lime, or alkalis diffolved in nitrous acid, without effervefcence or red vapours.

MERCURY diffolved in aqua regia yields corrofive fublimate ; whence, as well as from other confiderations, the author concludes, that the muriatic acid exifts in corrofive fublimate, in a dephlogifticated ftate. He has fince given (Journ. de Phyf. Mai 1785.) a very beautiful and fimple proof of the fame pofition. By only adding the dephlogifticated acid to a nitrous folution of mercury, he obtains corrofive fublimate. Nitrous folutions of mercury become more corrofive, as they are more deprived of phlogifton.

FROM thefe, and fome other experiments, the author thinks himfelf fully entitled to conclude, that the corrofive quality of metallic falts depends on their attraction for phlogifton.

P. 129.] NOTWITHSTANDING the ftrong attraction of the acid of fugar for lime, there are cafes in which it will not fhow its prefence. We have an inftance of this important practical obfervation in Mr Scheele's and our author's analyfis of the calculus. The former perceiving no precipitation to take place on the addition of acid of fugar, immediately concluded, that there was no lime prefent ; but the latter having often obferved, that a third fubftance fuperadded

to

to two already united, inſtead of effecting a ſeparation, enters into cloſe combination with them, ſuſpected, that this might be the caſe here, eſpecially as he knew, that the ſaccharine acid contains an unctuous matter, though of great ſubtilty. And upon burning ſome cal‑ culus to aſhes, obtained a ſubſtance which exhibited the moſt unequivocal marks of calcareous earth. (Stockh. Tranſact. vol. xxxvii. p. 333.) Hence we learn, how deſirable it is in chemiſtry to be poſſeſſed of more than a ſingle teſt, as it is called, of different ſubſtances.

P. 139. l. 21.] THIS perſon probably is Mr Hermbſtadt of Berlin; for we have a paper by him on this ſubject in Crell's Neueſt. Entdeck. part 9. p.6. It is obvious to ſuſpect, that the vegetable acids of ſu‑ gar and tartar, at leaſt, and perhaps of vinegar, are, at bottom, one and the ſame, only modified by ſome addition, rather accidental than eſſential. This ſuſpi‑ cion is favoured, not only by a reſemblance in ſenſible qualities; but alſo, by the production of one or other of theſe acids, according to the different circumſtances of a body, as in the ſeveral ſtages of fermentation. But ſuch conſiderations are, by no means, fitted to de‑ cide any chemical queſtion; they can only ſerve to ſuggeſt proper experiments. Accordingly, Mr Hermb‑ ſtadt attempted ſuch as were likely to decide the que‑ ſtion: one part of acid of tartar, treated with four parts of nitrous acid, (of which the ſpecific gravity was to water as 41 : 28,) as in the preparation of faccharine acid, yielded fome cryftals like thofe of this latter acid, but only in the proportion of $\frac{1}{18}$ in one experiment, and about $\frac{1}{9}$ in another.

But upon adding four ounces of fmoking nitrous acid to fix drachms of acid of tartar, and abftracting it with a brifk fire, he obtained four drachms and two fcruples of columnar cryftals, which produced, in a great number of experiments, the fame effects as the acid from fugar, and, in many refpects, different from that of tartar. Though fuch numerous proofs of coincidence fcarce leave any doubt, yet it is ftrange, that Mr Hermbftadt fhould neglect what may be confidered as the *experimentum crucis*, the precipitation of a folution of gypfum. He promifed, indeed, more experiments; but I have not been able to find them, either in the continuation of Crell's Journal, or any other work. It is however to be remembered, that Bergman treated the acid of tartar with nitrous acid, without obtaining any acid of fugar, [Opufc. vol. i.] Nor is it to be forgotten, that two vegetable or animal acids are very frequently prefent in the fame compound, as in the cafe of fugar of milk and microcofmic falt. But we can furely fcarce fuppofe, that fo large a proportion of acid of fugar fhould be accidentally prefent.

Mr Westrumb, another very intelligent German chemift, obtained four drachms, two fcruples of acid of fugar, from an ounce of tartar, treated with nitrous acid. His method of proceeding is worth mentioning.

To

To an ounce of tartar, he added two ounces of weak nitrous acid, and diſſolved it by means of a gentle heat. The liquor was then expoſed to evaporation in the ſun's rays, and, in ſome days, he obſerved cryſtals of nitre formed, amounting to two drachms, five grains ; when it would yield no more of theſe, two ounces of ſtrong nitrous acid were added to the acid and viſcid reſiduum ; when the phial had ſtood a ſhort time, red vapours began to ariſe ; the addition of ſtrong nitrous acid was repeated, as long as the liquor retained any viſcidity, or any red vapours aroſe, in which four ounces of nitrous acid were conſumed in all, and the quantity of ſaccharine acid obtained was four drachms, two ſcruples.

P. 158.] INSTEAD of obtaining phoſphoric acid by the tedious and waſteful method of combuſtion in the air, I ſhould think chemiſts would procure it by de-compoſing phoſphorus with nitrous acid, as Mr Lavoi-ſier directs, Mem. Paris, _ann._ 1780, p. 349. & ſeq. Nothing is required to procure the acid in a ſtate of as great purity, as by combuſtion, and with the great-eſt eaſe and expedition, but a prudent management of the fire.

P. 165.] THE proceſſes mentioned in this page, ſug-geſt what has been ſought by ſo many chemiſts, an unexceptionable method of preparing Pruſſian or phlogiſticated alkali ; all that remains to be done after

the

the acid has been once obtained, is to faturate it with an alkali or with an abforbent earth, which, if we may judge from feveral inftances mentioned in the text, feems to anfwer equally well. But a fhorter procefs occurred, in confequence of the difcovery of the nature of phlogifticated alkali, to Mr Scheele, and much about the fame time to Mr Weftrumb.

THE method of the latter is as follows : he faturates pure vegetable alkali, by frequently boiling it with well-wafhed Pruffian blue. He then boils the filtered liquor with white lead, in order to feparate any ful-phureous or phlogiftic particles that may happen to adhere to it. He then adds vinegar, which when it has been diftilled in tin-veffels, occafions the precipi-tation of a white matter in great abundance ; but not a particle of blue is feen to fall. He then, in con-formity with Scopoli's advice, expofes the liquor to the fun's rays, and keeps it in that fituation as long as any red precipitate is obferved to feparate. Upon this the lixivium is filtered, and then mixed with a double quantity of highly rectified fpirit of wine, which throws down the proper falt of the *lixivium sanguinis* in the form of fhining flocculi ; they are to be feparated by means of the filter, and all the faline matter foluble in fpirit of wine is to be extracted by that menftruum. The folution of the falt in water, is of a bright yellow colour ; does not fhew the leaft veftige of iron upon the addition of an acid ; precipi-

tates

tates that metal of a beautiful blue colour, copper red, &c.

Mr Scheele's method is far lefs complicated. He extracts, as before, Pruffian blue, with perfectly cauftic fixed alkali, and then mixing highly rectified fpirit of wine with the filtered liquor, he obtains the falt in the form of flocculi. Mr Scheele adds, that he is thoroughly convinced of the inefficacy of every other method of purifying the *lixivium fanguinis ;* for if the yellow folution be properly boiled with muriatic or vitriolic acid, Pruffian blue will always be feparated. The falt obtained by the procefs juft defcribed, is not liable to alteration in the open air ; for the iron holds the tinging acid in clofer union with the alkali, and fixes it fo that it cannot be diflodged by the aerial acid, which otherwife would happen, was it not combined with iron or fome other metal in the tinging lixivium.

P. 187. § 42. Magnefia.] Mr Butini of Geneva (*Nouv. Obferv. et Recherch. fur la Magnefie. A Geneve.*) having lately publifhed feveral curious obfervations on magnefia, which have not, as far as I know, been laid before the Englifh reader, I am tempted to give a fhort account of them, although they are not fo immediately connected with the doctrine of attractions. He was not acquainted with the differtation of Bergman on this earth ; but he nearly agrees with him, in faying that an ounce of diftilled

water

water is capable of diffolving a grain, or at moft $1\frac{1}{4}$
grain of magnefia, whereas the fame quantity of a-
erated water takes up thirteen grains.   He found
that magnefia does not at firft diffolve in aerated
water, but decompofes it by attracting the fixed air ;
when once faturated, it diffolves without decompofition; in its ordinary ftate, therefore, this earth is not
faturated with aerial acid, the alkali ufed for its precipitation not fupplying it with a fufficient quantity.
Mr Butini determines by exact experiments, that faturated magnefia contains $\frac{1}{10}$ of the aerial acid, more
than in its ordinary ftate.   The folution in aerated
water in the proportion of 1 : 64, becomes turbid in a
temperature of 158° of Fahrenheit.   But one of the
moft remarkable among his obfervations, is, that water
may be over-faturated with magnefia, and yet pafs
through the filter, and feem clear.   Such a folution
is obtained by immediately filtering the water in which
Epfom falt has been decompofed ; if it be heated to
68° Fahrenheit, (which may be done in the palm of
the hand), it lets fall its earth, which is rediffolved
when the liquor cools to about 59°.   This is a very
amufing experiment.

   THE fpontaneous cryftallization is alfo a new and
curious phænomenon.   It will abandon the water,
even when that is not faturated, in order to arrange its
integrant parts into regular forms.   The cryftals are
hemifpherical *matrices*, confifting of needles, from the
length of half a line, to that of five or fix lines,
                                                which

which are tranfparent hexaedral prifms, terminated by an hexagonal plane.

WHEN it is made to cryftallize in a temperature of 59°—62°, two kinds of cryftals are formed, *viz.* groups of needles and folitary *blocks*, of which the fhape is not exactly defined, though it is to be referred to that of an hexaedral prifm, terminated by an hexagonal pyramid. In a heat of 39° to 41°, nothing but blocks appear; and again from the 73° to the 77°, the needles only are formed.

BY repeated or violent calcination, magnefia lofes its property of eafy folubility in acids. Its particles, without acquiring a greater degree of mutual cohefion, gain a remarkable hardnefs, whence they become capable of fcratching fteel, &c. Water does not diffolve above $\frac{1}{5792}$ of calcined magnefia, nor does the folution yield any cryftals. During calcination, this earth emits a phofphoric light, and adheres with great tenacity to cold bodies, it alfo prefents that appearance of fluidity, which is remarked in gypfum. Calcined magnefia, expofed to an atmofphere of aerial acid, or left in a veffel covered only with a piece of loofe paper, does not recover its fixed air.

NEUTRAL falts heighten the folvent power of water, while alkalis diminifh it.

To thefe experiments of Mr Butini, let me be allowed to fubjoin fome of another author, not lefs remarkable, though they relate to another part of the chemical hiftory of magnefia, *viz.* its combination with acids.

MR

Mᴿ Quatremere D'Isjonval (Coll. des Memoires, p. 207.) gives an account of his having obtained permanent compounds, with magnefia and nitrous
acid; and what is more extraordinary ftill, with muriatic
acid likewife. To obtain this effeᴄt with the firft,
he precipitated purified Epfom falt, diffolved in cold
water, with fixed vegetable alkali ; he then faturated
the magnefia with pure nitrous acid, and evaporated
the folution, which at firft yielded nitre, on account
of fome alkali carried down by the precipitate. After
this was feparated, he rediffolved the faline magma,
and evaporated it again : Thefe operations he repeated
two or three times, till fome rudiments of cryftals
appeared, which being rediffolved, afforded, on evaporation, cryftals that had a ftronger tendency to ef-
florefce than deliquiate, even in a moift place. They
have the form of four-fided prifms, truncated acutely.
With muriatic acid he proceeds much in the fame
manner, taking care to faturate the acid completely.

He fubjoins two cautions of importance for the
more certain and fpeedy produᴄtion of thefe cryftals :
1. That not above half the magnefia fhould be precipitated, for he is afraid of the prefence of calcareous
earth; and 2. That the magnefia fhould be diffolved
in the acid, while it is yet in the tender form of a
precipitate.

I was unwilling to with-hold thefe curious obfervations of a chemift who had carried away the prize,
when Bergman was his competitor ; but, whatever
authority

authority this may add to his name, I think that he who shall perufe his writings, will be careful how he gives entire credit to his affertions, before he has repeated his experiments, though it muft be confeffed, that they receive fome authenticity from a letter of Mr de Morveau (p. 222.) to whom Mr D'Isjonval fent fpecimens of his cryftals.

P. 198. *paragraph* B.]. I MUST confefs myfelf ignorant of any good reafon for believing *phlogifticated, foul,* or *corrupted* air, to be a modification of vital air. Mr Kirwan's reafons for fuppofing it to be aerial acid, combined with more phlogifton, convey to me no fort of conviction. Mr Cavendifh has thrown a ray of light upon this obfcure fubftance, as I have already mentioned, and unlefs his rational conjecture fhould be ripened into a difcovery, it is better to own our entire ignorance of the nature of this elaftic fluid, than to content ourfelves with any of the explanations that have yet been offered.

P. 199. & feq. *paragraph B.*]. THE connection between nitrous acid and vital air now begins to appear in a very different light. To fuppofe that thefe two fubftances were but modifications of one and the fame, was both natural and allowable, when vital air was firft procured from nitre ; but when it appeared, in the progrefs of enquiry, that fo many other bodies, free from all fufpicion of any mixture of nitrous acid, were

were found to yield the fame fluid, the opinion could be no longer tenable, nor is it, by any means, conformable to the ufual feverity of the author's logic. The experiments of Mr Cavendifh and Mr Watt fhew, that the common office of nitrous acid and other fubftances, is merely to dephlogifticate water. The latter, who made an attempt to recover the nitrous acid, found, upon procuring vital air from this acid and earths, that, however thoroughly the acid and earth might be dephlogifticated, the acid always became highly phlogifticated after the procefs. (Ph. Tr. vol. lxxiv. p. 338.). He found, moreover, in one experiment, that thirty-fix ounces meafure of vital air were produced, and only five grains of weak nitrous acid miffing ; and in another, thirty four grains weight of the fame air were produced, with the lofs of only two grains of *real* acid, p. 343.

FURTHER, when vital air is obtained from vitriolic falts, vitriolic acid air appears, at the fame time, even when the falts are not known to contain any phlogiftic matter, p. 344.

P. 208. & feq. *paragraph C.*] IT is now no longer probable, either that Mr Kirwan's or Mr Scheele's opinion will be confirmed. Both muft give way to the difcovery of Mr Cavendifh, concerning the conftituent parts of water. By fome experiments made in the fummer of 1781, and read before the Royal Society in 1784, he found, that, upon

firing

firing together inflammable and vital air in clofe veffels, they were condenfed into water. Other chemifts,
both at home and abroad, have now amply confirmed
this unexpected obfervation, as Dr Prieftley, S Lau-
driani at Milan, and Mr Lavoifier at Paris, who has
ufed very large quantities in his experiments, but has
fhamefully attempted to appropriate the difcovery to
himfelf; and he is accordingly mentioned in many
foreign journals as the firft difcoverer. Dr Prieftley
found, that after inflammable and vital air had been
deflagrated together, and the veffel had cooled to the
temperature of the atmofphere, as much mercury or
water, in whichever of thefe liquids the mouth was
immerfed, entered, as was fufficient to fill it within
$\frac{1}{200}$ part of its contents: moreover, when the moifture
adhering to the glafs was wiped off with a piece of
fponge paper, firft carefully weighed, it was found exactly, or very nearly, equal to the airs employed. (Mr
Watt, Ph Tranf. vol. lxxiv. p 332.). This difcovery
is fo much the more to be admired, as no hints had
been thrown out by any other author which could
lead to it, nor could it have been furmifed by any
analogical reafoning. It promifes, however, to furnifh explanations of many of the obfcureft operations,
both in art and nature. Thus the generation of vital air, the difappearance of vital and nitrous air,
when mixed together, the production of vital air by
vegetables, the diminution of the air in the combuftion of fulphur, phofphorus, &c. are now no longer

ger phænomena that require, for their explanation, hypothefes uncountenanced or contradicted by experiments. But Mr Cavendifh's difcovery leads to much wider views. It fuggefts new experiments on the increafe of weight of calcined metals, a problem ftill remaining to be folved, notwithftanding fo many late attempts. The firft object of thofe who fhall now labour on this fubject, fhould be to afcertain, whether a quantity of water, equal to the difference of weight, is generated during calcination. The general principle is equally applicable to volcanos and to ftatical phyfiology ; for it is now obvious to fufpect that the exhalation from the lungs is not thrown off in that form by animals, but rather generated by the mixture of the dephlogifticated part of the common air with phlogifton. But this is not the proper place to indulge in fuch fpeculations.

To offer any arguments againft Mr Scheele's doctrine of the compofition of heat, would be now fuperfluous, fince Mr Kirwan (Notes to the Treatife on Air and Fire) and Fontana (Opufc. Litt. à S. Adolph. Murray) have abundantly confuted it. Bergman, himfelf, notwithftanding he has adapted feveral explanations to it, feems to acknowledge at laft, that it is liable to infurmountable objections. One alone is fufficient : that lofs of weight, which muft enfue if the vital air and phlogifton pafs off through the veffels in the form of heat, is not obferved. But it is a proof of Mr Scheele's acutenefs, that he firft perceived

ved the neceffity of fome fubftance containing vital air and phlogifton in fo many chemical experiments.

MR KIRWAN'S explanation, is equally inadmiffible ; for Mr Cavendifh has fhewn, in the moft fatisfactory manner, that no fixed air is generated by the mixture of nitrous and vital air, any more than in the explofion of inflammable air ; or at leaft, if any be generated, it is fo fmall a quantity, " as to elude the " niceft teft we have." (Ph. Tr. vol. lxxiv. p. 121. 122.; and 172. 173.)

P. 211. l. 3. and 4.] MR SENEBIER (Recherches fur l'air inflammable) has found that inflammable air actually does change vital air in a length of time. By keeping thefe fubftances over water, he obferved that a diminution took place, and that the refiduum did not undergo any alteration on the addi. tion of nitrous air. From what Mr Kirwan fays, (Ph. Tr. vol. lxxiv. p. 168.) it feems that Dr Prieftley has made obfervations to the fame purport ; for he tells us, that the Doctor has difcovered, fince his laft publication, that inflammable and dephlogifticated air will unite.

P. 223. l. 5. & feq.] PRECIPITATE per fe, and red precipitate, are foluble in marine acid, and during the folution, nothing is difengaged, but a great heat is produced, as in the flaking of quicklime ; the

salt

falt which is obtained by reducing the folution to cryftals, is corrofive fublimate.

THIS obfervation does not coincide with the experiment of Bergman, who affirms, that calcined mercury is reduced by digeftion in muriatic acid. I cannot
guefs how he could have made his experiment; for
whenever I have added precipitate per fe to muriatic
acid, I have always obferved folution to take place
with the production of much heat, and have obtained
cryftals of corrofive fublimate on refrigeration. It is
true indeed, that when red precipitate is employed, a
black powder, confifting of mercury, feparates; but
this mercury is not any of the calx revivified; it feparates during the time of folution, becaufe the muriatic diffolves no mercury *that is not combined with*
vital air; and this mercury happened to be mixed
with the calx; fo, if we take notice of what paffes
in the preparation of red precipitate, it will be feen
that firft nitrous air, and afterwards red precipitate
pafs over, long before the mercury rifes. May not
Bergman have ufed this precipitate, and thus been
led into a miftake? He knew not how to explain the
reduction. Mr Kirwan (Phil. Tranf. vol. lxxiv.
p. 159) fays that the reduction is owing to the expulfion of the fixed air from the mercurial calx; which
fixed air, at the moment of its expulfion, is decompofed, leaving its phlogifton to the mercury, which is
thereby revived. But this explanation is inadmiffible,
                                                    fince

ſince no reduction takes place, as I have already obſerved. (*Pelletier J. de Phyſ. Mai*, 1785.)

P. 253. l. 11. & ſeq.] DR PRIESTLEY ſays, that
the iron is *ſuperphlogiſticated*, and has lately aſſured us,
[Phil. Tranſ. vol. lxxiii.] that he has frequently repeated the experiment with the ſame reſult. I fear,
that neither Prieſtley nor Bergman have examined
the ſtate of the iron, otherwiſe than ſuperficially.
Perhaps, on a more accurate examination, ſomething
might appear that would lead to an explanation of the
phænomenon.

P. 257. l. 4. & ſeq.] CRELL adviſes (Chem. An-
nal. St. i. p. 94.) to uſe in this experiment another
equal and ſimilar veſſel, containing a quantity of water
A, of equal weight, and at the ſame temperature
(32° Fahr.) with the ice B; let the two veſſels, he
adds, be placed near one another, and obſerve how
long a time A requires to attain the temperature of the
room. Moreover, obſerve how long a time B requires, 1ſt, to melt, and, 2d, to attain the temperature of the room. Next, multiply the degree of heat
acquired by A, by the time B 1ſt, took to melt, and, 2d,
to attain the temperature of the atmoſphere. Then,
ſuppoſing the degrees of the thermometer to correſpond to equal acceſſions of the matter of heat, we
may ſay, as much heat, (*i. e.* ſo many particles of fire),

as is fufficient to raife the thermometer to a certain heat, has fuch a weight.

I HAVE heard, that one of thofe Philofophers who believe heat to be merely a quality, has made the experiment propofed by Bergman, with this ftrange refult, that, during the melting of ice, a diminution of *abfolute* gravity takes place. If my information be right, I fhould think fuch a conclufion could prove nothing, but either the difficulty of making the experiment, without fuffering fome of the vapours to efcape, or elfe the carelefsnefs of the obferver.

P. 258. 259. 260.] MR LAVOISIER and De la Place have lately given the following table of fpecific heats :

Common water, - -	1,
Sheet iron, - -	0,109985
Glafs without lead, or cryftal, -	0,1929
Mercury, - -	0,029
Quicklime, - -	0,21689
Mixture of water and quicklime, in the proportion of 9 to 16, -	0,334597
Oil of vitriol, (fpec. grav. 1,87058,)	0,60362
Mixture of this oil with water, in the proportion of 4 to 5, -	0,663102
Nitrous acid, (not fmoking, fpecific gravity, 1,29895,) •	0,661391

Mixture

Mixture of this acid with quicklime, in
   the proportion of $9\frac{1}{3}$ to 1,   -   0,61895
Mixture of 1 part of nitre with 8 of
   water,   -   -   0,8167

THE ingenious machine, contrived for the purpofe
of meafuring the fpecific heat of bodies, by thefe
gentlemen, confifts principally of three cavities inclofed in one another. The outermoft is filled with
pounded ice, which is defigned to intercept the influence of the atmofphere upon the ice contained in the
middle cavity. This latter is the real fubject of the
experiment, and the quantity of it melted by the bodies which are placed in the innermoft cavity, gives
the proportion of their fpecific heats, which are evidently as the quantity of water obtained. But there
is reafon to fear left this ingenious and promifing contrivance fhould fail of anfwering the end propofed ;
for Mr Wedgewood, who repeated fome of the experiments, obtained the moft oppofite refults, when
the fame body was heated to the fame degree. His
difappointment arofe chiefly from two caufes; the firft
of which was the abforption of the water by the
pores of the ice ; for, though the French Philofophers pretend, that no error can arife from this caufe,
fince the ice has already imbibed as much water as it
can, yet it is obvious, that, as the precife degree of
force with which the pounded ice is preffed together
cannot be adjufted, more or lefs water will be fucked
up,

up, as it forms a lefs or more compact body. The other difficulty feems ftill more infuperable. Mr Wedgewood found, that the two proceffes of freezing and thawing were going on at the fame time :—whether it be, that water in the ftate of vapour is liable to freeze in a higher temperature than when liquid, as Mr Wedgewood conjectures, or the ordinary abforption of heat by evaporation alone produces the effect, without any fuch difpofition in the vapour. At all events, the prefent table, as well as thofe which Mr Lavoifier and De la Place promife, muft be received with diffidence, till thefe impediments to accuracy be removed.

P. 263. *note.*] THE fubftance mentioned in this note, is called black wadd ; by Da Cofta, *ochra friabilis nigra.* Its remarkable property of taking fire when mixed with oil, was firft difcovered accidentally at a painter's in Derby in 1752. Mr Wedgewood has lately (Phil. Tranf. vol. lxxiii. p. 284.) publifhed an analyfis of it. He finds it to confift of infoluble earth, chiefly micaceous, of lead, iron, and manganefe.

22 parts contain

of earth about	2
of lead	1
of iron	$9\frac{1}{2}$
of manganefe	$9\frac{1}{2}$
	22

P. 268.

P. 268. 269.] THIS analyſis of charcoal differs exceedingly from that of Dr Prieſtley. He obtained not the leaſt particle of aerial acid from charcoal, when it had been perfectly well burned ; but the whole quantity was converted, by means of burning lens, into inflammable air, except a very inconſiderable portion of aſhes. (Phil. Tranſ. vol. lxxiii. p. 411.)

P. 273.] IT is now completely aſcertained by the experiments of Mr Hutchins at Hudſon's bay, that the freezing point of mercury correſponds to the 40th degree below o of Fahrenheit's thermometer ; therefore the numbers in the text muſt be much reduced. The other data of Dr Crawford, viz. the converſion of vital into fixed or phlogiſticated air, have, by no means, the certainty of eſtabliſhed principles. We muſt now look for the origin of the heat, in the condenſation of vital air into water by phlogiſton. And thus we have Mr Cavendiſh's diſcovery extended to another of the moſt obſcure and familiar phænomena in nature.

P. 278.] BEFORE the publication of Mr Cavendiſh's paper on air, (Phil. Tranſ. vol. lxxiv. p. 119. & ſeq.) Mr Kirwan ſeems to have almoſt ſucceeded in perſuading chemiſts, that fixed air is generated in phlogiſtic proceſſes, by the union of vital air with phlogiſton. Others had thrown it out long before as a probable ſuppoſition ; but Mr Kirwan was, I think, the

the firſt. who, by collecting and arranging the numerous facts publiſhed by different authors, gave the
opinion a great degree of plauſibility. Still, however, the complete proof, from unequivocal, analytical, and fynthetical experiments, was wanting, and many of the moſt important caſes of phlogiſtication gave
no fort of countenance to the ſuppoſition.

MR CAVENDISH, by a rigorous examination of the
arguments, has fairly reduced them within a very
narrow compaſs. In the firſt place, he juſtly ob-
ſerves, that all experiments, in which any organiſed
bodies are employed, muſt be ſet aſide. He is inclined,
likewiſe, to conſider the experiments with the electric
ſpark as equivocal; being of opinion, that when turnſol
is uſed, the aerial acid may ariſe from the burning of
this vegetable matter; and, in the caſe of lime water,
from ſome impurity in the tube, or elſe from ſome inflammable matter in the lime.

THERE remain then but four caſes; the calcination
of metals, the combuſtion of ſulphur or phoſphorus,
the mixture of nitrous air, and the exploſion of inflammable with vital air.

NEITHER in the combuſtion of ſulphur and phofphorus, nor in the exploſion of inflammable air, has
any veſtige of aerial acid been perceived. That this
is alſo the caſe in the mixture of nitrous air, Mr Cavendiſh has clearly ſhown, contrary to the general ſuppoſition, as I have already obſerved.

WE

WE have, therefore, left, for the fupport of fo important and extenfive a doctrine, only the calcination
of metals : And in examining thefe experiments, made
with care and clofe veffels, we do not find any evidence of the generation of fixed air. Mr Lavoifier
and Dr Prieftley found none in the air in which they
had formed their experiments ; and, if it be faid, that
the metallic calces abforbed it, it is anfwered, that
none has been extracted from calces fo prepared. It
is indeed true, that metallic calces prepared in open
veffels, or fuch as have lain long expofed to the atmofphere, contain aerial acid; but here the atmofphere
is an evident fource from which it might arife.

MR KIRWAN endeavours to remove thefe objections, by adducing feveral experiments, which fhow,
that aerial acid exifts in very fparing quantities in the
atmofphere, and by reprefenting it as improbable, that
metals, during their calcination, fhould attract it ;
becaufe lime, expofed to a red heat ever fo long,
does not regain any. He infifts upon what he had formerly advanced, that the revivification of certain mercurial calces, and the production of vital air, is owing
to the decompofition of the aerial acid contained in
the calces.

HE thinks his opinion ftrongly corroborated by an
experiment of Mr Laffone, in which filings of zinc having been digefted with cauftic alkali, an effervefcence
was obferved on the addition of an acid ; but Mr Cavendifh fuppofes, that the effervefcence arofe, not

from

from the expulſion of aerial acid from the alkali, but of inflammable air from the incompletely diſſolved zinc.

WITH reſpect to the experiment of Dr Prieſtley, mentioned in the text, Mr Cavendiſh relates an experiment, which ſeems to render it probable that the aerial acid proceeds from plumbago and other impurities contained in the iron : 500 grains of red precipitate, mixed with 1000 of iron filings, yielded 7800 grain meaſures of aerial acid ; and 2400, partly of vital, and partly inflammable air. But 500 grains of the ſame red precipitate yielded 9200 of aerial acid, and 4200 of indifferent vital air ; when they were mixed the plumbago and other impurities, which were the reſiduum of 1000 grains of iron filings, diſſolved in diluted vitriolic acid. Hence, as more aerial acid was produced when the red precipitate was mixed only with the impurities, than with the iron filings themſelves, it ſhould ſeem to follow, that its production was owing, not to the iron, but to the plumbago, which is known to contain a great deal of it. As, however, it was found, that the iron filings were mixed with $\frac{1}{13}$ of their weight of braſs, and, as more fixed air was produced, than the plumbago uſually contained in 1000 grains of iron can ſupply, (which is Mr Kirwan's objection), Mr Cavendiſh candidly acknowledges, that the experiment ought to be repeated in a more accurate matter.

Mʀ

MR KIRWAN, befides, infifts upon the diminution of common acid by the electric fpark, as the moft convincing argument in favour of his opinion. An experiment, too, in which Dr Prieftley having amalgamated lead with mercury, obtained aerial acid from the black powder into which the lead was converted, feems really favourable to him.

SUCH is the ftate of a doctrine which fo widely influences the theory of chemiftry. Every one muft acknowledge, that much ftronger proofs muft be adduced before it can be received as a fundamental propofition. Now, when we know that vital air and phlogifton conftitute water, and that this will fufficiently account for the diminution of the air, it is no longer a neceffary hypothefis: And this very difcovery feems to give the opinion a general appearance of improbability, which the few facts that yet remain to countenance it, do not, by any means, outweigh ; nay, it is probable, that thefe facts, as they are more narrowly fcrutinifed, will put on a different afpect, as has already happened in the mixture of nitrous and vital or common air. May not fome fuch experiments as I have propofed above, for exploring the nature of phlogifticated air, ferve to fhow whether aerial acid is really a compound of phlogifton and fome other fubftance ?

P. 285. Æther.] THE reader will furely excufe me if I digrefs a little from the fubject of attractions, and the text of the author, in order to give him a
                                                          fhort

short account of some late experiments of Mr Scheele
on this substance. They are to be found in the
Stockholm Transactions, part iii. for the year 1782.
Mr Scheele has not indeed yet been able fully to
clear up the obscure theory of the generation of æ-
ther; but his experiments have led him to some new
views.

WHEN vitriolic æther is prepared in a large retort,
and a brisk heat is applied towards the last, volatile
sulphureous acid is obtained together with vinegar,
but no vestige of aerial acid:—when to an ounce of
pounded manganese, half that quantity of vitriolic acid,
and an ounce of strong spirit of wine was added, he
got both vinegar and aerial acid, and found in the retort vitriolated manganese, without any excess of acid.
He found that vitriolic and muriatic acids are consti-
tuent parts of their respective æthers, but in exceedingly small proportions.   Besides zinc, antimony, and
tin, by the intervention of which, it is well known
that muriatic æther may be made, he obtained æther
by a solution of bismuth in aqua regia, evaporated to
the thickness of a syrup, and of crocus martis (iron
filings will not answer the purpose) in muriatic acid.
He also obtained æther by saturating spirit of wine
with fluor acid air, and adding manganese   No acetous æther is to be obtained by the process of the
Count de Lauraguais, notwithstanding almost all chemical writers, except Poerner, have admitted that
process as an effectual one : Mr Scheele, however,
found

found a way of preparing acetous æther, and that, in greater quantity than any other kind of æther (this obfervation is furely important in practice) by adding to ftrong vinegar a little vitriolic, nitrous, muriatic, or fluor acid. Acetous æther is more eafy of decompofition than any other. No æther could be procured with phofpi.oric acid, nor any with falt of benzoin alone, though the latter yields fome with the help of muriatic acid. Neither did his trials fucceed with acid of tartar, of lemons, of borax, of amber, and feveral compound falts.

CONCERNING the theory, he obferves, that though it may feem, that fome fubftance, which has an attraction for the phlogifton of fpirit of wine muft be brought into action ; yet this fuppofition can fcarce be applied to the acetous æther, or that of benzoin, or to the acids of fluor and fea-falt. But granting that thefe fubftances actually do attract phlogifton, how is the *oil of fpirit of wine* itfelf, or the æther, feparated from the water with which it was before united ? Perhaps this phænomenon may be explained in the fame way as the feparation of fulphur from hepatic air : this latter is foluble like fpirit of wine in water, and confifts of phlogifton, fulphur, and the principle of heat. On the acceffion of a body that feparates the phlogifton of this air, the principle of heat efcapes, and the fulphur is precipitated. In applying this fuppofition, Mr Scheele obferves, that manganefe has a ftrong attraction for the inflammable

principle

principle when an acid acts upon it. This metallic calx, fpirit of wine, and vitriolic or muriatic acid be- ing added together, the former attracts part of its in- flammable principle from the fpirit, whence the heat (which is fo confiderable, that this mixture boiled of itfelf) efcapes, and the fubtile oil or the æther is fepa- rated from the water. The portion of acid in high- ly-rectified æther, is very infignificant, though it can be made obvious. The fmall quantity of acetous and aerial acid, which he has obferved in fome diftillations, proceeds from the decompofition of a fmall portion of æther ; for it is very probable, that the oil of fpirit of wine confifts of acetous acid and the principle of inflammability.

P. 318. Siderite.] This fuppofed new metal, fi- derite, [*Hydrofyrum*, *Waffereifein*] has been reduced by the firft difcoverer, Mr Meyer of Stettin, to a mere compofition of iron and phofphoric acid. The reader will be beft pleafed with his own reflections on the fubject. " It is, fays he, (Crell's Chemifche Anna- " len, B. i. St. 3.) but too eafy to fall into miftakes " in chemiftry ; a truth daily confirmed by the num- " ber of difputed and contradictory experiments, and, " of which, I myfelf probably furnifh a frefh ex- " ample. For my new metal, obtained from caft i- " ron, that was run from a marfhy ore, of which I " have treated in the obfervations of the Berlin So-
" ciety

" ciety * is, in all probability, neither more nor less
" than iron combined with pholphoric acid. My rea-
" fons for this opinion are the following. I diffolved
" fome of my fuppofed new metal in vitriolic acid,
" taking care to ufe more acid than was neceffary for
" the folution. I obtained in the retort a gray
" powder, and fome fulphur appeared on its neck.
" Upon diffolving the gray powder, and evaporating
" the folution, I obtained a thick brown lixivium,
" in which, when it had flood ftill for a confiderable
" time, fome cryftals of true martial vitriol fhot:
" The remainder of the ley fhewed the fame phæ-
" nomena, as thofe I have noticed in the above men-
" tioned papers. Hence it was too apparent, that
" this metallic fubftance contained a large quantity of
" iron. But with what could it be combined? I
" could think of nothing but phofphoric acid.

    " HAVING poured a little water upon 20 grains of
" iron that had been fluxed with inflammable matter,
" and afterwards forged, I dropped into it a little phof-
" phoric acid, procured by the burning of phofphorus,
" and applied heat. The acid attacked the iron, and
" what was diffolved formed a gray powder. I ad-
" ded, by degrees, acid enough to diffolve all the i-
" ron, and then left it to dry by fpontaneous evapo-
" ration. The gray powder, when dry, weighed
" $55\frac{1}{2}$ grains. Upon trying to fufe half a drachm of
" this with 20 grains of glafs of borax, I found that
              " it

* B. ii. p. 334. ; and B. iii. p. 380.

" it did not flow well : nor when 20 grains more
" were added, did it run into complete fusion. I
" found in the glafs particles of metal that were melt-
" ed into grains not perfectly round, which were al-
" fo found to be very brittle, fufed with difficulty
" under the blow-pipe, and were converted into
" fcoriæ. The magnet had but little effect on them,
" attracting only fome fmall particles. Upon the re-
" maining earth I poured oil of vitriol, diluted with
" an equal quantity of water; it was left to dry, and
" then diffolved in a very fmall quantity of water,
" and filtered it. Upon mixing this folution with wa-
" ter, it became milk-white, and there fell down a
" confiderable quantity of white earth, in appear-
" ance like the earth of fiderite. I have not yet
" been able to repeat and continue thefe experi-
" ments ; but have no doubt of their being confirm-
" ed ; and, in that cafe, I muft alter the title of my
" effays; but I hope the effays are not without their
" ufe. What a plentiful fource of phofphoric acid
" would be opened to us, if it were but eafy to fe-
" parate ? The clofe combination of this acid with
" iron, would alfo be remarkable.

" BERGMAN has adopted my water-iron (Waffer-
" eifen) as a new metal, under the title of fiderum."

WHAT Mr Meyer gathers from thefe two experiments, is confirmed by Affeffor Klaproth of Berlin,
who, by a remarka le coi idence, came to the very
fame conclufion, without any communication with Mr
Meyer.

Mcyer. He did not attempt to eftablifh his opinion by analytical experiments, as he conceived that it would be difficult to feparate the iron and acid, either by phlogifton or any other way. He found, however, the artificial compound of phofphoric acid and iron, to agree in its properties with the *calx fideri alba* obtained by Bergman and Meyer from cold-fhort iron. Native Pruffian blue contains this combination in much larger proportion.

* * * * *

Since the firft publication of this differtation in 1775, befides many alterations which totally change the difpofition of it, not lefs than nine new columns have been added, which, if all the rectangles were filled up, make $9 \times 50 = 450$ new rectangles. I have no doubt but that an equal or a greater number of additions will hereafter be made in an equal number of years. Even fince the publication of the third volumn of the *Opufcula* in 1783, two, or perhaps three fubftances have been difcovered which will claim a place on the table of elective attractions; thefe are what Mr Scheele confiders as the acid inherent in tungftein, or *lapis ponderofus*, the metal which Meffrs Luyart obtained from tungftein; for Wolfram only differs from it in being combined with iron and manganefe, and the acid of the filk-worm, and fome other infects, defcribed by Mr Chauffier in the Dijon Memoires for 1783. Little, I prefume, is as yet known
concerning

concerning the elective attractions of thefe fubftances.
Mr Scheele tells us, that the tungftein acid, as he fuppofes it to be, when combined with volatile alkali, decompofes nitrated lime, by a double elective attraction,
and regenerates tungftein. He adds, that it produces
no change on folutions of alum or vitriolated lime, but
it decompofes acetated ponderous earth, the precipitate being quite infoluble in water ; that vitriolated
iron, zinc, and copper, nitrated lead, filver, and mercury, with acetated lead, are precipitated white, and
muriated tin-blue ; and that corrofive fublimate is
not changed. All this was well known to Bergman ;
yet he has made no ufe of it, perhaps wifely judging
that the experiments of Mr Scheele, as well as his own,
could not be much relied upon, on account of the
fmall quantity of matter they had to work upon : And,
in fact, Meffrs Luyart found, that what the Swedifh
chemifts had taken for the acid of tungftein, was a triple falt, containing, befides the fubftance furnifhed by
the ftone, fome of the acid and alkali employed in extracting and faturating it.

* * * * *

Since the notes on p. 96. *note*, and fome other paf-
fages, were printed, I have had the fatisfaction of fee-
ing a paper lately read by Mr Cavendifh before the
Royal Society, and containing experiments which
deeply affect fome of the general theories mentioned
as well in the differtation as the notes. This paper,
joined

joined to that which I have so frequently quoted, tends
to clear up more obscurities in this branch of chemistry, than all the other facts and theories with which
I am acquainted put together ; and with whatever
modesty and simplicity his experiments may be related
by the author, they ought to be accepted by those who
have been perplexed by the endless doubts and difficulties that occur in all that has been written on aeriform substances, as great and important discoveries.

I HAVE already mentioned (p. 326.) the idea, that
Mr Cavendish had thrown out on the nature of phlo·
gisticated air, and observed, what was very obvious
from the consideration of that philosopher's experiments, that the matter might be fully ascertained, by
treating this elastic fluid with vital air. Mr Cavendish
did not fail to pursue the path which thus lay open before him. He mixed phlogisticated and vital air together, and passed the electric spark through the mixture. In these trials, a diminution of bulk always was
observed, insomuch that when five parts of vital air
were added to three of common air, almost the whole
disappeared. Moreover, by continuing his experiments, he discovered, that an acid liquor was produced, and that this acid was the nitrous. Thus, was
his conjecture concerning the constitution of phlogisticated air fully confirmed ; and as, in the experiments
already published, he had shown, that vital air is the
same thing as water deprived of phlogiston ; so, in the
present case, the addition of vital air is equivalent to
the addition of water.

BUT

BUT his paper contains other experiments on the subject of aerial acid, that are equally interesting. Mr Kirwan, so far from admitting the validity of Mr Cavendish's objections to these experiments in which the elastic spark is made to pass through common or vital air, confined by a solution of litmus or of lime, retains them as the best arguments in favour of that opinion which he has espoused. It was therefore desirable to see how far experiment would countenance them.

WHEN the electric spark was taken in small portions of common air, confined by a solution of litmus, the liquor was turned red. This had been observed before ; but it was not before known, that, by continuing the sparks, the solution becomes quite clear and transparent ; so, however, it is : besides, half the air disappears, and, by the addition of lime-water, it is reduced $\frac{1}{5}$ more. It is, therefore, unquestionably true, that the litmus suffers a decomposition, loses its purple colour, and yields fixed air ; but there is nothing in any of these, or of the following experiments, which favours the opinion of air being diminished by means of phlogiston communicated by means of the electric spark.

WHEN lime-water was used instead of litmus, not the least cloud was observed ; the air was reduced to $\frac{2}{3}$ of its original bulk ; whereas, by phlogistication *,

it

* To shew the impropriety of the former erroneous phraseology, is always among the first consequences of new discoveries ; for, as the expression depends on the opinion that is formed of phænomena,

it lofes but $\frac{1}{5}$. When the fpark was taken in common or vital air not quite pure, no cloud was perceived, nor even when fome aerial acid was introduced ; but when, befides aerial acid, pure volatile alkali was added, a brown fediment immediately appeared. Hence, it is evident, as Mr Cavendifh obferves, that the calcareous earth muft have been faturated by fome acid, (*viz.* the nitrous), which was generated during the experiment ; but, on the addition of pure volatile alkali, four powers came into action, the nitrous acid uniting with the volatile alkali, and the aerial acid with the lime. The brown colour of the fediment might be owing to fome quickfilver being diffolved. It is, moreover, undeniable, that, if any aerial acid had been generated, it would have precipitated fome of the earth, till a fufficient quantity of the other acid was produced to diffolve the whole.

THUS then it appears, that what in the opinion of its warmeft maintainer are the moft convincing proofs of the doctrine in queftion, are no proofs at all in its favour.

na, the former way of fpeaking will fcarce ever be adapted to the new notions. By altering the meaning of the prefent phrafe, it may perhaps be retained, fince the vital part of the air does actually receive phlogifton; but it feems to me to exprefs better, that the remaining elaftic fluid has received an addition of phlogifton, by which its nature has been changed: whence, there will arife this inconvenience, that paft writers will have ufed the fame expreffion with one meaning, and future writers will ufe it with another. I think, therefore, that confufion will be beft avoided, by fubftituting a new one.

favour. I never experienced greater furprife than on reading thefe experiments of Mr Cavendifh. The precipitation of lime, in thefe circumftances, was fo generally believed, that no one had fcrupled to affume it as a certain ground of reafoning. Speculative men cannot learn, from a more ftriking inftance, how neceffary it is to begin with a ftrict examination of facts.

But whence arifes the aerial acid that appears in thofe phlogiftic proceffes in which animals are concerned? We find it not only in the air that has been refpired by animals provided with lungs, but Mr Scheele detected it in air in which he had kept infects, and Mr Achard in common and vital air that had been injected into the cellular tiffue of animals. If it does not proceed from any change of vital air, what remains but that it muft be thrown off in fubftance? I would, therefore, propofe this as a proper fubject of experiment, not becaufe I think it calculated to decide the chemical queftion concerning the conftitution of the aerial acid, but becaufe it is a curious phyfiological problem. I know but one fact that has any immediate connection with the folution of this problem, and that is contained in Mr Achard's paper on artificial emphyfemas. He always found a large portion of aerial acid in inflammable air that had been forced into the cellular tiffue of animals. If it was not contained beforehand in the inflammable air, and little or no common air was introduced along with it, which I think altogether unlikely, the refult feems very much to favour the opinion

368 NOTES.

nion that fuppofes aerial acid to be an animal emana-
tion.

I KNOW not whether I fhall efcape cenfure for
dwelling fo long, both here and before, on the queftion concerning the aerial acid ; but I prefume, with
fome confidence, that no one who is capable of perceiving its extenfive influence on the theory of fo many chemical operations, will blame me with much feverity. That the author of the differtation confidered
it as a queftion of the utmoft importance, appears
from his accurate ftatement of the feveral theories, and
of the objections that may be made to them. *Gaudemus
interea*, fays he, *quæftionem eo reduétam fuiffe, ut certi-
tudo diu defiderari nequit;* and accordingly, I think,
we can now determine, with tolerable certainty, the
merit of thofe theories, as well as explain the nature of phlogiftic proceffes.

ON this fubject, I have only to add, that in column
36. *vital air* ought to be placed above *nitrous acid.*

P. 188.] I FORGOT to obferve, that ponderous
earth had now been found, in more places than one,
combined with aerial acid. The author himfelf received, a little while before hi death, a fpecimen of
this combination, which was tranfmitted to him from
this country.

EXPLANATION

Explanation of the TABLE of Double Elective Attractions.

The firſt Forty take place in the Humid Way.

Scheme 1. repreſents the decompoſition of vitriolated vegetable alkali by ponderous earth, the vitriolated ponderous earth falling down inſoluble, and the pure alkali remaining in the water.

Scheme 2. ſhews that lime produces no ſuch effect.

Scheme 3. the decompoſition of muriated foſſil alkali or ſea-ſalt by pure vegetable alkali, the new compound and the diſengaged alkali both remaining in the liquor.

Scheme 4. the decompoſition of muriated lime (fixed ammoniac) by pure foſſil alkali, the lime falling to the bottom.

Scheme 5. Lime produces no change in a ſolution of muriated fixed alkali.

Scheme 6. the decompoſition of vitriolated magneſia (Epſom ſalt) by pure fixed alkali, the pure magneſia nefia falling to the bottom, and the new compound being diffolved.

SCHEME 7. the decompofition of corrofive fublimate by pure vegetable alkali, the mercurial calx falling down, and the new compound (digeftive falt) being diffolved.

SCHEME 8. the decompofition of vitriolated iron (green vitriol)by lime, both the metallic calx and the new compound (gypfum) being precipitated.

SCHEMES 9. 10. 11. Thefe fchemes are admirably explained in the ninth fection ; fee alfo p. 170. The ninth reprefents the partial decompofition of vitriolated vegetable alkali by the nitrous acid : the tenth and eleventh that of nitre and digeftive falt by the acid of tartar.

N. B. There is an error in the tenth Scheme of the original plate, which, as well as many others, is corrected in the Tables annexed to this Tranflation.

SCHEME 12. fhews, that, on the contrary, the acid of tartar does not, in any refpect, change muriated foffil alkali (fea-falt).

SCHEME 13. the decompofition of borax by nitrous acid, the new compound being diffolved, and the acid of borax appearing in a folid form.

SCHEME

SCHEME 14. the decompofition of vitriolated lime
(gypfum) by acid of fugar, the faccharated lime falling down infoluble.

SCHEME 15. fhews the decompofition of vitriolated
magnefia by fluor acid : but fee p. 187.

SCHEME 16. fhews the decompofition of nitrated
lime by vitriolic acid, the gypfum falling down.

SCHEME 17. fhews the decompofition of white arfenic by dephlogifticated marine acid, the acid of arfenic
appearing in a folid form.

SCHEME 18. fhews that zinc precipitates filver from
volatile alkali.

SCHEME 19. fhews, that faline liver of fulphur is decompofed by the acetous acid, the fulphur being precipitated.
N. B. This Scheme includes many others, a fortiori.

SCHEME 20. fhews the decompofition of calcareous
liver of fulphur by vitriolic acid, both gypfum and fulphur falling down.

SCHEME 21. fhews, that when folutions of vitriolated vegetable alkali and muriated lime are mixed, a
double decompofition takes place, the digeftive falt being diffolved, and the gypfum precipitated.

SCHEME

SCHEME 22. fhews, that when vitriolated vegetable alkali in folution is added to muriated lead (plumbum corneum), a double decompofition takes place, the fea-falt being diffolved, and the vitriolated lead precipitated.

SCHEME 23. fhews, that when muriated vegetable alkali and vitriolated lime are mixed together, no double decompofition takes place.

SCHEME 24. fhews, that when vitriolated volatile alkali and muriated mercury are added together, a double decompofition takes place, the muriated volatile alkali (fal ammoniac) being diffolved, and the vitriolated mercury precipitated.

SCHEME 25. fhews, that when nitrated terra ponderofa and oxalited volatile alkali are added together, an exchange of principles takes place, the oxalited earth falling down, and the nitrous ammoniac being diffolved.

SCHEME 26. fhews the double decompofition of nitrated filver and common falt, the new neutral falt (nitrated foffil alkali) being diffolved, and the muriated filver (argentum corneum) falling down.

SCHEME 27. fhews, that when tartar and nitrated mercury are mixed, they fuffer a mutual decompofition,

tion, the nitre being diffolved, and the tartarized mercury precipitated.

Scheme 28. fhews, that when borax and nitrated mercury are mixed, a double decompofition takes p'ace, the boraxated mercury falling down, and the nitrated foffil alkali diffolved.

Scheme 29. fhews, that when muriated magnefia and acetated filver are mixed, an exchange of principles takes place, the acetated magnefia being diffolved, and the muriated filver precipitated.

Scheme 30. fhews, that when vitriolated filver and muriated lead are mixed, a double exchange happens, and both muriated filver and vitriolated lead fall down infoluble.

Scheme 31. fhews, that when nitrated filver and muriated copper are mixed together, a double decompofition takes place, the nitrated copper being diffolved, and the muriated filver falling down.

Scheme 32. fhews, that when common falt and aerated vegetable alkali are mixed, a double exchange takes place, the digeftive falt (muriated vegetable alkali) and the aerated foffil alkali being both diffolved.

Scheme

SCHEME 33. ſhews, that when corroſive ſublimate and aerated vegetable alkali are mixed, a double de-compoſition happens, the muriated vegetable alkali be-ing diſſolved, and the aerated mercury precipitated.

SCHEME 34. ſhews, that when nitrated lead and aerated foſſil alkali are mixed, a double exchange is ef-fected, the aerated lead falling down, and the nitrated foſſil alkali being diſſolved.

SCHEME 35. ſhews, that when vitriolated magne-ſia and aerated fixed alkali are mixed together, a double exchange takes place, the vitriolated alkali being diſſol-ved, and the aerated magneſia falling down.

SCHEME 36. ſhews, that when muriated lime and aerated volatile alkali are mixed together, a double ex-change takes place, the muriated volatile alkali (ſal ammoniac) being diſſolved, and the aerated lime fall-ing down.

SCHEME 37. ſhews, that when aerated volatile alkali is added to a tincture of pure vegetable alkali, the vegetable alkali combines with the aerial acid, and falls down, while the volatile alkali combines with the al-cohol.

SCHEME 38. ſhews, that when to nitrated ſilver copper is added, a double exchange takes place, the

phlogiſton

phlogifton of the copper unites with the filver, which falls down, and the nitrous acid with the copper, which is diffolved.

SCHEME 39. fhews, that if iron be added to cop- per diffolved in vitriolic acid, the phlogifton combines with the calx of copper, which is precipitated, and the vitriolic acid with the calx of iron, which is diffolved.

SCHEME 40. fhews, that when faline liver of ful- phur is mixed with acetated lead, a double exchange takes place, the lead falling down with the fulphur, and the acetated alkali being diffolved.

N. B. I cannot undertake to affign the reafon why thofe cafes of double attraction which are effected by the tinging acid, in combination with alkalis, and, as I fuppofe, with abforbent earths, at leaft with lime, are omitted in the prefent Table. The nature of the ope- ration could not be unknown to the author, becaufe he mentions it exprefsly, p. 302. I fee, indeed, that he has omitted others that were known to him ; nay, that he mentions in this very differtation ; whether it was that the reader could eafily infer them from thofe cafes that are reprefented in the Schemes, or whether they were doubtful, or of fmall importance in practice : but none of thefe confiderations apply to the attrac- tions in queftion ; for they feem to be well afcertained as to their theory, and are undoubtedly of daily occur- rence in the practice of chemiftry.

IT

IT remains, that the deficiency be, in fome meafure, fupplied; and I cannot do this better than by the following table, which I have extracted from the author's effay on metallic precipitates. The firft column fhews the colour of the precipitate; the fecond its quantity, from 100 grains of the regulus, diffolved in fome acid menftruum.

	Colour.	Weight.
Gold,	yellow,	{ not perfectly precipitated.
Platina,		no precipitation.
Silver,	dark yellow,	145.
Mercury,	{ whitifh at firft; turns yellow when dry,	} apt to be rediffolved.
Lead,	white.	
Copper,	{ greenifh yellow; blackifh red when dry,	530.
Iron,	deep blue,	590.
Tin,	white,	250.
Bifmuth,	yellowifh,	180.
Nickle,	{ yellow; dark brown when dry,	250.
Arfenic,	white,	180.
Cobalt,	{ bluifh red; brownifh red when dry,	142.
Zinc,	{ white; citron yellow when dry,	495.
Antimony,	white,	138.
Manganefe,	{ at firft bluifh, then yellowifh blue, laftly dark green.	

THE precipitates of tin, bifmuth, nickle, antimony, and manganefe, generally have their colours alter- ed by the admixture of particles of Pruffian blue, from the iron that is prefent. The ponderous earth is likewife precipitated by the combination of the tinging acid and alkali, (Preface to the Sciagraphia, and Withering, *ibid.*). Hence I imagined, that vitriolated ponderous earth might poffibly be decompounded by the fame powers. Having re- duced a fmall quantity to a fine powder, I digefted it in the funfhine for feveral hours in a folution of Pruffian alkali, prepared according to Mr Scheele's method. After the fupernatant liquor was poured off, and the refiduum edulcorated, nitrous acid was added to it, which I fuppofed would expel the tinging acid from the earth, if any fuch union had taken place, and dif- folve the latter ; but upon dropping fome vitriolic acid into the nitrous, no precipitation was obfervable, which fhews that it held no ponderous earth in folution. Whether this decompofition may be accomplifhed by applying a ftronger heat, or following other methods, I know not.

THIS experiment led me to doubt, whether the Pruffian alkali was capable of decompounding certain other compounds of acids and metals, befides the fo- lution of platina. If any would refift its action, I thought vitriolated and muriated lead, and perhaps filver, to be the moft likely ; particularly the former, from the conjecture that Bergman has thrown out in

the

the preface to his *Sciagraphia,* concerning the identity of the calx of lead and ponderous earth. I was, moreover, unable to recollect any thing satisfactory on this point in authors. Bergman himself, (Opusc. Diff. xxiii. § 5. letter E), says only, that " lead is " precipitated of a white colour from its solution in " nitrous acid by aerated and caustic fossil alkali, as " well as by Prussian alkali." I attempted, therefore, to satisfy myself, by repeating the foregoing experiment with vitriolated lead, except that I happened to use the acetous acid instead of the nitrous. But I now observed a white precipitate, both on the addition of vitriolic and muriatic acid, which seems to show, that a double decomposition had actually taken place. However, farther enquiry is necessary before any thing certain is determined; and I purpose to examine the several metallic salts above mentioned, when I have some tolerable convenience for making chemical experiments.

SCHEME 41. If vitriolated vegetable alkali be distilled with phosphoric acid, it will be decompounded; the vitriolic acid will be driven over, and the phosphorated alkali will remain at the bottom.

SCHEME 42. If common salt be subjected to distillation with nitrous acid, the marine acid will rise, and the nitrated fossil alkali remain in the retort.

SCHEME

SCHEME 43. If common falt be fubjected to diftillation with acid of arfenic, the marine acid will rife, and the arfenicated foffil alkali remain in the retort.

SCHEME 44. If muriated volatile alkali (fal ammoniac) be diftilled with vitriolic acid, the marine acid will rife, and the vitriolated volatile alkali will likewife be fublimed.

SCHEME 45. If marine acid be diftilled with the black calx of manganefe, it will be dephlogifticated and rife, the white calx remaining fixed.

SCHEME 46. If fal ammoniac be treated with quicklime in diftilling veffels, the pure volatile alkali will arife, and the muriated lime remain.

SCHEME 47. If fulphurated mercury (cinnabar) be fubjected to diftillation with iron, the mercury will rife in its metallic form, and the fulphurated iron remain.

SCHEME 48. If arfenicated fixed alkali be fublimed with inflammable matter, the arfenic will rife in a reguline form, and the alkali will remain?

SCHEME 49. If phofphorated volatile alkali be fublimed with inflammable matter, it will be decompofed, and the volatile alkali and phofphorus will both arife?

SCHEME

SCHEME 50. fhews, that muriated lime is not decompounded by being expofed to diftillation with volatile alkali.

SCHEME 51. fhews, that fluorated lime is not decompounded by being expofed to fufion in a crucible with fixed vegetable alkali.

SCHEME 52. fhews, that when an alloy of gold and filver is fufed with fulphur, the fulphur combines with the filver, and leaves the gold free.

SCHEME 53. fhews, that when fulphurated lead and iron are fufed together, the fulphur unites with the iron, and leaves the lead free.

SCHEME 54. fhews, that when copper is fufed with a combination of faline liver of fulphur and filver, the copper unites with the liver of fulphur, and the filver is feparated.

SCHEME 55. fhews, that when nitre and muriatic acid are diftilled together, the nitrous acid attracts the phlogifton of the marine acid, and quits the vegetable alkali.

SCHEME 56. fhews, that when nitre and white arfenic are diftilled together, the nitrous acid at-

tracts

tracts the phlogifton, and rifes, while the arfenical acid combines with the vegetable alkali.

SCHEME 57. fhews, that when common falt and white arfenic are fubjected to diftillation together, no change is effected.

SCHEME 58. fhews, that when corrofive fublimate is fubjected to diftillation with regulus of antimony, the calx of the mercury unites with the phlogifton of the antimony, and rifes, while the muriatic acid combines with the calx of antimony, and rifes in like manner.

SCHEME 59. fhews, that when vitriolated vegetable alkali is expofed to fublimation with arfenicated volatile alkali, the vitriolic acid combines with the volatile alkali, and rifes, while the arfenical acid combines with the vegetable alkali, and remains fixed.

SCHEME 60. fhews, that when nitrated vegetable alkali and vitriolated volatile alkali are expofed to fublimation, the nitrous acid rifes, combined with the volatile alkali, while the vitriolated vegetable alkali remains fixed.

SCHEME 61. fhews, that when muriated foffil alkali and vitriolated mercury are fublimed together, the muriatic acid combines with the mercury, and
rifes,

rifes, while the vitriolated foffil alkali remains at the bottom.

SCHEME 62. fhews, that when muriated volatile alkali and aerated lime are expofed to fublimation together, the aerial acid and volatile alkali rife combined, while the muriated lime remains fixed.

SCHEME 63. fhews, that when aerated vegetable alkali and fluorated lime are fufed together, the acid of fluor combines with the vegetable alkali, and the aerial acid with the lime, both remaining fixed.

SCHEME 64. fhews, that when an alloy of gold and copper is fufed with fulphurated antimony, the fulphur unites with the copper, and the gold with the antimony, both remaining fixed.

E M E N-

EMENDANDA.

P. 41. l. 16. add *vitriolated tartar, nitre, and some other salts, are thus thrown down.*

P. 66. l. the laſt, ſtrike out *or not.*

P. 68. l. 12. for *a* read *n.*

P. 107. l. 13. for *acids,* read *metals.*

P. 170. l. 5. after *manner,* inſert *the acids of.*

P. 182. l. 17. read the words, *which come next,* immediately after *ſea-ſalt.*

P. 236. l. 14. inſert : between the two laſt quantities of the equation.

P. 256. l. 17. after *ſpecific,* add *heats.*

N. B. There are a few *literal errors,* which the context will at once direct the reader how to correct : And if there be a few wrong ſtrokes yet remaining in Table III. they may be corrected from the correſponding Table of words.

APPENDIX I

A NOTE ON THE IDENTITY OF THE TRANSLATOR

The title page of the translation does not give the translator's name but merely says that it has been 'translated from the Latin by the Translator of Spallanzani's Dissertations'. The English translation of the Abbé Spallanzani's *Dissertations relative to the Natural History of Animals and Vegetables* (London, 1784) is also anonymous. Both translations were formerly attributed to Edmund Cullen of Dublin, who did indeed publish a translation of many of Bergman's *Physical and Chemical Essays* (2 vols., London, 1784; 2nd edition 1788; vol. iii, Edinburgh, 1791). This translation does not include the *Dissertation on Elective Attractions*.

After the Translator's Preface, however, there is the following Advertisement:

> The great distance of the Translator obliged the publisher to call in the assistance of another person to superintend the press. That person, being unacquainted with the translator's design of adding annotations at the end of the volumes, was induced to add a few inconsiderable observations, such as his recollection could furnish, during an hasty perusal of the MS. He, moreover, thought, that the addition of M. de Morveau's notes would be acceptable to the English reader. He accordingly has selected such as afford any new views of the facts related by the Author, or supply any experiment by which the text is illustrated. He has omitted several that did not seem to contain much information. That no blame might be imputed to the Translator, on account of these additions, he has been careful to distinguish those of M. de Morveau by his name; his own by the letter B. Those marked C. belong to the Translator; who has been obliged, on account of some domestic interruption, to defer the greater part of his remarks to the end of the second volume; which will appear without delay. The Author's notes and references are without a signature.

That the author of the notes signed B in Cullen's translation of some of the other Essays was Thomas Beddoes, and that he went on to translate the *Dissertation on Elective Attractions*, is difficult to establish on stylistic grounds. The latter, and the translation of Spallanzani's *Dissertations*, do seem better stylistically and more accurate than Cullen's work; but their style naturally tends to follow their originals to some extent. Nevertheless, the evidence provided by the English translation of Scheele's *Chemical Essays*, published in London in 1786 by the same publisher, is almost conclusive. The Preface to that is signed by Beddoes, and in it he writes:

> . . . having so lately annexed to BERGMANN'S DISSERTATION ON ELECTIVE ATTRACTIONS, much of what it would have been proper to observe concerning Mr. Scheele's experiments and deductions, as far as I am acquainted with it, I chuse rather to refer to that publication, than repeat the same things in the present (pp ix–x).

The mis-spelling of Bergman's name as Bergmann is also found on the title page of the present work.

The story is made clear in the *Memoirs of the Life of Thomas Beddoes* (London, 1811) by J. E. Stock, whose preface shows that he knew Beddoes and had acquired a great deal of information from his other friends and from unpublished manuscripts. Stock takes it as a well-known and admitted fact that Beddoes was the translator of both Spallanzani's *Dissertations* and Bergman's *Dissertation on Elective Attractions*, and argues that the notes signed B in Cullen's translation were written by Beddoes not only from their internal evidence but 'from the faint traditional recollection of the gentleman who first pointed them out to my notice; who was nearly his contemporary, and was, like him, a pupil of Sheldon'. We may therefore take it as certain that Beddoes was the author of the present translation of Bergman's *Dissertation*.

Beddoes was born in 1760 at Shifnal in Shropshire and after graduating at Pembroke College, Oxford, studied medicine under John Sheldon in London, and at Edinburgh. He took the degree of Doctor of Medicine at Oxford in 1786 and was Reader in Chemistry there from 1788 to 1792. His lectures were exceedingly well attended, though the appointment was not very profitable for him financially. He had visited Guyton de Morveau and Lavoisier in France.

In 1796 he published his novel on the evils of drink, *The History of Isaac Jenkins*, over 40,000 copies of which were sold before the end of the year. Beddoes worked hard for some years for the establishment of a Pneumatic Institution for the treatment of diseases by the inhalation of different gases. He gained the support of James Watt, Josiah Wedgwood, and other associates of the Lunar Society of Birmingham, and the Institution was eventually set up in 1798 at Clifton, Bristol. Humphry Davy was its first Superintendent, but left in 1801 to join the Royal Institution. Belief in the healing power of gases had already waned, and the Institution became an ordinary hospital. Beddoes gave up his connection with it in 1807, the year before his death. He had written on social and political subjects as well as on chemistry and medicine. His wife was the daughter of Richard Lovell Edgeworth, and the sister of Maria Edgeworth, the novelist. Thomas Lovell Beddoes (1803–49), the dramatist and poet, was their son.

APPENDIX II

Glossary of Names of Substances and Other Terms Occurring in the Text, Tables or Notes and their Modern Equivalents.

(It should be remembered that many eighteenth-century terms have no exact modern equivalents because the substances to which they referred are now differently identified.)

Acescency.	Souring, sourness.
Acetated lead.	Lead acetate.
Acetated ponderous earth.	Barium acetate.
Acetous acid.	Acetic acid.
Acetous aether.	Acetic ester, ethyl acetate.
Acid of amber.	Succinic acid.
Acid of ants.	Formic acid.
Acid of arsenic.	Arsenic acid.
Acid of benzoin.	Benzoic acid.
Acid of borax.	Boric acid.
Acid of fat.	The acid obtained by the distillation of animal fats such as suet, probably a mixture of stearic acid, oleic acid and other high fatty acids. (See Nicholson, *First Principles of Chemistry*, 3rd edition, London, 1796, p. 353.)
Acid of fluor	Hydrofluoric acid.
Acid of lemon.	Citric acid.
Acid of milk.	Lactic acid.
Acid of nitre phlogisticated.	Oxides of nitrogen.
Acid of phosphorus.	Phosphoric acid.
Acid of Prussian Blue.	Prussic acid.
Acid of salt, acid of sea salt.	Hydrochloric acid.
Acid of sorrel.	Oxalic acid.
Acid of sugar.	Oxalic acid. Bergman distinguished between acid of sorrel and acid of sugar, but Scheele showed that they were the same.

Acid of sugar of milk. — Mucic acid.

Acid of tartar. — Tartaric acid.

Acid of vitriol phlogisticated. — Sulphur dioxide.

Acidum perlatum. — See page 161. This acid, obtained from human urine, was thought to be an additional component of microcosmic salt, but was found by Scheele in 1785 to be no different from phosphoric acid.

Aerated alkali. — Potassium or sodium carbonate.

Aerated calx (of lead). (Page 298). — Lead carbonate.

Aerated fixed alkali. — Potassium carbonate or sodium carbonate.

Aerated lime. — Calcium carbonate.

Aerated mercury. — Carbonate of mercury.

Aerated silver. — Silver carbonate.

Aerated vegetable alkali. — Potassium carbonate.

Aerated volatile alkali. — Ammonium carbonate.

Aerial acid. — Carbon dioxide.

Aeriform marine acid. — Hydrogen chloride gas.

Aeriform nitrous acid. — Gaseous oxides of nitrogen.

Alkali of tartar by deliquescence. — Potassium hydroxide.

Aqua regia. — A mixture of nitric and hydrochloric acids.

Argenteum corneum. — Horn silver, silver chloride.

Argillaceous earth. — Clay.

Arsenicated vegetable alkali. — Potassium arsenate.

Assa dulcis. — Benzoin.

Butter of antimony. — The deliquescent chloride of antimony.

Butter of arsenic. — Chloride of arsenic.

Calcareous earth. — Lime.

Calcareous hepar. — Calcium sulphide, or polysulphides.

Calx (plural *calces*). — Generally what is now called the oxide of a metal, which was regarded in Bergman's day as a simple substance from which the metal was formed by combination with phlogiston. The Latin word also means lime, as does the French equivalent *chaux*, and the same symbol is used for lime and for a metallic calx.

Calx of mercury precipitated per se.	Mercuric oxide.
Calx of quicksilver.	Oxide of mercury.
Caustic (of alkali).	In aqueous solution, as opposed to solid or gaseous state.
Caustic fixed alkali.	Sodium or potassium hydroxide.
Corrosive sublimate.	Mercuric chloride.
Crocus martis.	Ferric oxide.
Cubic nitre.	Sodium nitrate.
Dephlogisticated air.	Oxygen.
Dephlogisticated marine acid or muriatic acid.	Chlorine.
Dephlogisticated sea salt.	Chlorine.
Depurated.	Purified.
Digestive salt.	Potassium chloride.
Divaricating.	Branching, diverging.
Earth of alum.	Aluminium oxide.
Earth of hartshorn.	Hartshorn regarded as a chemical substance.
Earth of ivory.	Ivory regarded as a chemical substance.
Edulcoration.	Washing to remove water-soluble particles.
Elastic fluid.	Gas.
Elixation.	Boiling, in water or alkaline solution.
Essential oil.	The lighter and more volatile oils obtained from aromatic plant substances by pressure or distillation with water.
Expressed oil.	Heavier oil, obtained by pressure.
Fixed fossil alkali.	Sodium hydroxide.
Fixed mineral alkali.	Sodium hydroxide.
Fixed vegetable alkali.	Potassium hydroxide.

Flaming nitre.	Ammonium nitrate.
Flowers of zinc.	Zinc oxide.
Fluorated alkali.	Potassium or sodium fluoride.
Foeculant matter.	Foul matter.
Formicated lime.	Calcium formate.
Glauber's salt.	Sodium sulphate.
Green vitriol.	Ferrous sulphate.
Hepar.	Liver, as in liver of sulphur.
Hepatic air.	Hydrogen sulphide.
Inflammable air.	Hydrogen.
Lamellae.	Thin plates.
Lapis calaminaris.	Calamine, zinc carbonate.
Lapis infernalis.	Silver nitrate.
Lapis ponderosus.	Tungsten oxide.
Liquor of flints.	Flint dissolved in a solution of potassium or sodium hydroxide in water, i.e. a solution of potassium or sodium silicate.
Liquor silicum.	Liquor of flints, q.v.
Lixivium.	Alkaline solution, lye.
Lixivium sanguinis.	Potassium ferrocyanide.
Lunar vitriol.	Silver sulphate.
Magnesia alba.	Magnesia, magnesium oxide.
Magnesia nigra.	Manganese dioxide.
Marine acid.	Hydrochloric acid.
Marine air.	Hydrogen chloride gas.
Martial vitriol.	Sulphate of iron.
Matter of heat.	The fluid supposed to be responsible for the phenomena of heat.
Menstruum.	Solvent.
Metallic calces.	See *calx*.
Microcosmic salt.	Sodium ammonium phosphate, $Na(NH_4)HPO_4$.
Minium.	Lead oxide, Pb_3O_4.
Minimum dephlogisticated.	Litharge, PbO.
Molecule.	Small particle, but not in the technical sense in which the word has

	since been used to mean a group of atoms.
Muriated cobalt.	Cobalt chloride
Muriated copper.	Copper chloride
Muriated fossil alkali.	Sodium chloride.
Muriated lime.	Calcium chloride.
Muriated magnesia.	Magnesium chloride.
Muriated tin.	Stannous chloride.
Muriated volatile alkali.	Ammonium chloride.
Muriatic acid.	Hydrochloric acid (Latin *muria*, brine).
Muriatic air.	Hydrogen chloride gas.
Nitrated copper.	Copper nitrate.
Nitrated lime.	Calcium nitrate.
Nitrated lead.	Lead nitrate.
Nitrated magnesia.	Magnesium nitrate.
Nitrated mercury.	Nitrate of mercury.
Nitrated silver.	Silver nitrate.
Nitrated volatile alkali.	Ammonium nitrate.
Nitrous acid.	Nitric acid. Nitrous acid in the modern sense had not yet been separately described.
Nitrous air.	Gaseous oxides of nitrogen.
Nitrum flammans.	Ammonium nitrate.
Ochre.	Red oxide of iron.
Orpiment.	Sulphide of arsenic.
Oxaline acid.	Oxalic acid.
Oxalited lime.	Calcium oxalate.
Pabulum of fire.	The food of fire, i.e. oxygen.
Perlate acid.	Acidum perlatum, which in fact was phosphoric acid.
Perlate salt.	Microcosmic salt, sodium ammonium phosphate.
Phlegm.	Water, or watery distillate.
Phlogisticated acid of fluor.	Hydrofluoric acid.
Phlogisticated air.	Air in which something has burnt until there is no longer enough oxygen to allow it to continue to burn, i.e. mainly nitrogen.

Phlogisticated alkali. — Potassium ferrocyanide.

Phlogisticated nitrous acid. — Oxides of nitrogen.

Phlogisticated vitriolic acid. — Sulphur dioxide.

Phlogiston. — The principle of inflammability which is found in inflammable bodies and emitted when they burn or calcine.

Phosphorated alkali. — Potassium or sodium phosphate.

Phosphorated vegetable alkali. — Potassium phosphate.

Phosphorated lime. — Calcium phosphate.

Plumbum corneum. — Lead chloride.

Ponderous earth. — Barium oxide.

Powder of Algaroth. — Antimony oxychloride.

Prismatic nitre. — Potassium nitrate.

Pure (of alkalis). — In solid or gaseous form, as opposed to solution.

Pyrophorus. — A substance which grows hot on exposure to the air. Homberg's pyrophorus was obtained by heating alum with human excrement.

Quicksilver. — Mercury.

Quadrangular nitre. — Sodium nitrate.

Regulus (plural *reguli*). — The elementary form of a metal or semi-metal.

Regulus of antimony. — Metallic antimony.

Regulus of arsenic. — Elementary arsenic.

Regulus of copper. — Metallic copper.

Saccharated vegetable alkali. — Potassium oxalate. Bergman distinguishes this salt, formed from 'acid of sugar' and 'vegetable alkali', from the potassium salt of acid of sorrel, which is in fact also potassium oxalate.

Saccharum Saturni. — Sugar of lead, lead acetate.

Sal mirabile perlatum. — Microcosmic salt, sodium ammonium

	phosphate, incorrectly distinguished from ordinary microcosmic salt.
Sal perlatum.	Microcosmic salt.
Saline hepar.	Liver of sulphur, saline liver of sulphur.
Salt of Seignette.	Potassium sodium tartrate.
Salt of sorrel.	Potassium oxalate.
Sapo Helmontii.	Ammonium carbonate.
Secret sal ammoniac.	Ammonium sulphate. (On page 23 'secret' is misprinted as 'seret'.)
Sea salt.	Common salt, sodium chloride.
Sebaceous acid.	Acid of fat, i.e. the mixture of fatty acids obtained from suet.
Sedative salt.	Boric acid.
Sel vegetal.	Potassium tartrate.
Siderite.	Believed by Bergman to be a distinct metal, but really iron phosphide.
Silex.	Flint.
Siliceous earth.	Flint.
Sorrelled lime.	Calcium oxalate.
Spirit of wine.	Ethyl alcohol.
Succinated lime.	Calcium succinate.
Sugar of lead.	Lead acetate.
Tartar.	Potassium hydrogen tartrate.
Tartarized ponderous earth.	Barium tartrate.
Tartarized tartar.	Potassium tartrate.
Terra ponderosa.	Heavy earth, barium oxide.
Tincture of turnsole.	Litmus.
Tinging acid.	Prussic acid.
Tungstein acid.	Tungstic acid.
Turbith mineral.	Basic mercuric sulphate.
Turnsole.	Litmus.
Unctuous oil.	The heavier and less volatile vegetable oils.
Van Helmont's soap.	Ammonium carbonate precipitated with alcohol.
Vegetable salt.	Potassium tartrate.
Vital air.	Oxygen.

Vitriol of copper.	Copper sulphate.
Vitriol of mercury.	Sulphate of mercury.
Vitriolated calcareous earth.	Calcium sulphate (gypsum).
Vitriolated clay.	Aluminium sulphate.
Vitriolated cobalt.	Cobalt sulphate.
Vitriolated copper.	Copper sulphate.
Vitriolated fossil alkali.	Sodium sulphate.
Vitriolated iron.	Iron sulphate.
Vitriolated lead.	Lead sulphate.
Vitriolated lime.	Calcium sulphate.
Vitriolated magnesia.	Magnesium sulphate.
Vitriolated mineral alkali.	Sodium sulphate.
Vitriolated silver.	Silver sulphate.
Vitriolated tartar.	Potassium sulphate.
Vitriolated vegetable alkali.	Potassium sulphate.
Vitriolated volatile alkali.	Ammonium sulphate.
Vitriolic acid.	Sulphuric acid.
Vitriolic aether.	Ethyl ether.
Volatile alkali.	Ammonia.
Volatile hepar.	The product obtained by distilling sulphur with lime and ammonium chloride.

INDEX

[The names and spellings used in the index follow those of the Text. For their modern equivalents see Appendix II of the Introduction.—ED.]

Chemical Signs explained.

Acids.

1. + 🜍 vitriolic
2. + 🜍♄ phlogisticated
3. + 🜕 nitrous
4. + 🜕♄ phlogisticated
5. + 🜔 marine
6. + 🜔♃ dephlogisticated
7. ♀R Aqua regia
8. + 🜊 of fluor
9. ⚬🜔⚬ arsenic
10. + 🜕 borax
11. + ⊗ sugar
12. + 🜿 tartar
13. + ⊕ sorrel
14. + C lemon
15. + ⚬♀⚬ benzoin
16. + ∞ amber
17. + ⊘ sugar of milk
18. ‡ acetous distilled
19. + 🜂 milk
20. + ſ ants
21. + 🝁 fat
22. + 🜍 of phosphorus
23. + 🜍 perlatum
24. + 🝀 of prussian blue
25. 🝁 aerial

Alkalis.

26. ⊖Vp pure fixed vegetable
27. ⊖Mp pure fixed mineral
28. ⊖Λp pure volatile

Earths.

29. ⚥r pure ponderous
30. ⚥P pure calcareous. Lime
31. ⚥P pure magnesia
32. ∇r pure argillaceous
33. ⚟r pure siliceous
34. ∇ water
35. 🜁 vital air
36. 🜂 phlogiston
37. △ matter of heat
38. 🜍 sulphur
39. ⊖🜍 saline hepar
40. V̇ spirit of wine
41. ⚬⚬ Æther
42. ⚬⚬⚬ essential oil
43. ◎ unctuous oil

Metallic Calces.

44. ♀⊙ gold
45. ♀∞ platina
46. ♀☽ silver
47. ♀☿ mercury
48. ♀♄ lead
49. ♀♀ copper
50. ♀♂ iron
51. ♀♃ tin
52. ♀♄ bismuth
53. ♀🜨 nickle
54. ♀∞ arsenic
55. ♀R cobalt
56. ♀Θ zinc
57. ♀🝁 antimony
58. ♀🝆 manganese
59. ♀♂ siderite

Tab. I.

Tab. II.

Single, elective, Attractions.
In the Moist Way.

In the Dry Way.

Tab. III.

Single elective Attractions.
In the Moist Way.
In the Dry Way.

SINGLE ELECTIVE ATTRACTIONS.

In the Moist Way.

	1 VITRIOLIC ACID	2 PHLOGISTICATED VITRIOLIC ACID	3 NITROUS ACID	4 PHLOGISTICATED NITROUS ACID	5 MARINE ACID	6 DEPHLOGISTICATED MARINE ACID	7 AQUA REGIA	8 ACID OF FLUOR	9 ACID OF ARSENIC	10 ACID OF BORAX	11 ACID OF SUGAR	12 ACID OF TARTAR	13 ACID OF SORREL	14 ACID OF LEMON	15 ACID OF BENZOIN	16 ACID OF AMBER	17 ACID OF SUGAR OF MILK	18 ACETOUS ACID DISTILLED	19 ACID OF MILK	20 ACID OF ANTS
1	Pure ponderous earth	Pure ponderous earth	Pure vegetable alkali?	Pure vegetable alkali?	Pure vegetable alkali?	Pure vegetable alkali?	Pure vegetable alkali?	Pure vegetable alkali?	Lime	Lime	Lime	Lime	Lime	Lime	Lime	Pure ponderous earth	Lime	Pure ponderous earth	Pure ponderous earth	Pure ponderous earth
2	Pure vegetable alkali?	Pure fossil alkali?	Pure fossil alkali?	Pure fossil alkali?	Pure fossil alkali?	Pure fossil alkali?	Pure fossil alkali?	Pure fossil alkali?	Pure ponderous earth	Pure ponderous earth	Pure ponderous earth	Pure ponderous earth	Pure ponderous earth	Pure ponderous earth	Pure ponderous earth	Pure vegetable alkali?	Pure ponderous earth	Pure vegetable alkali?	Pure vegetable alkali?	Pure vegetable alkali?
3	Pure fossil alkali?	Pure fossil alkali	Pure fossil alkali	Pure fossil alkali?	Pure fossil alkali?	Pure fossil alkali?	Lime	Pure magnesia	Pure magnesia	Pure magnesia	Pure magnesia	Pure magnesia	Pure magnesia	Pure magnesia	Pure fossil alkali	Pure fossil alkali	Pure magnesia	Pure fossil alkali	Pure fossil alkali	Pure fossil alkali
4	Pure fossil alkali	Lime	Pure ponderous earth?	Pure ponderous earth	Lime	Lime	Pure magnesia	Pure vegetable alkali?	Pure vegetable alkali?	Pure vegetable alkali?	Pure vegetable alkali?	Pure vegetable alkali?	Pure vegetable alkali?	Pure vegetable alkali?	Pure volatile alkali	Pure volatile alkali	Pure vegetable alkali?	Pure volatile alkali	Pure volatile alkali	Pure fossil alkali
5	Lime	Pure magnesia	Lime	Lime	Pure magnesia	Lime	Lime	Pure fossil alkali	Pure fossil alkali	Pure fossil alkali	Pure fossil alkali	Pure fossil alkali	Pure fossil alkali	Pure fossil alkali	Lime	Pure fossil alkali	Pure fossil alkali	Lime	Lime	Pure volatile alkali
6	Pure magnesia	Pure volatile alkali	Pure magnesia	Pure magnesia	Pure volatile alkali	Pure magnesia	Pure magnesia	Pure volatile alkali	Pure volatile alkali	Pure volatile alkali	Pure volatile alkali	Pure volatile alkali	Pure volatile alkali	Pure volatile alkali	Pure volatile alkali	Pure magnesia	Pure volatile alkali	Lime	Lime	Lime
7	Pure volatile alkali	Pure clay	Pure volatile alkali	Pure volatile alkali	Pure clay	Pure volatile alkali	Pure volatile alkali	Pure clay	Pure clay	Pure clay	Pure clay	Pure clay	Pure clay	Pure clay	Pure volatile alkali	Pure magnesia	Pure volatile alkali	Pure magnesia	Pure magnesia	Pure magnesia
8	Pure clay	Pure clay	Pure clay	Pure clay	Pure clay	Pure clay	Pure clay	Calx of zinc	Pure clay	Pure clay	Pure clay	Pure clay	Pure clay	Pure clay	Pure clay	Pure clay	Pure clay	Pure clay	Pure clay	Pure clay
9	Calx of zinc	Calx of zinc	Calx of zinc	Calx of zinc	Calx of zinc	Calx of zinc	Calx of zinc	Calx of iron	Calx of zinc	Calx of zinc	Calx of zinc	Calx of zinc	Calx of zinc	Calx of zinc	Calx of zinc	Calx of zinc	Calx of zinc	Calx of zinc	Calx of zinc	Calx of zinc
10	Calx of iron	Calx of iron	Calx of iron	Calx of iron	Calx of iron	Calx of iron	Calx of iron	Calx of manganese	Calx of iron	Calx of iron	Calx of iron	Calx of iron	Calx of iron	Calx of iron	Calx of iron	Calx of iron	Calx of iron	Calx of iron	Calx of iron	Calx of iron
11	Calx of manganese	Calx of manganese	Calx of manganese	Calx of manganese	Calx of manganese	Calx of manganese	Calx of manganese	Calx of cobalt	Calx of manganese	Calx of manganese	Calx of manganese	Calx of manganese	Calx of manganese	Calx of manganese	Calx of manganese	Calx of manganese	Calx of manganese	Calx of manganese	Calx of manganese	Calx of manganese
12	Calx of cobalt	Calx of cobalt	Calx of cobalt	Calx of cobalt	Calx of cobalt	Calx of cobalt	Calx of cobalt	Calx of nickel	Calx of cobalt	Calx of cobalt	Calx of cobalt	Calx of cobalt	Calx of cobalt	Calx of cobalt	Calx of cobalt	Calx of cobalt	Calx of cobalt	Calx of cobalt	Calx of cobalt	Calx of cobalt
13	Calx of nickel	Calx of nickel	Calx of nickel	Calx of nickel	Calx of nickel	Calx of nickel	Calx of nickel	Calx of lead	Calx of nickel	Calx of nickel	Calx of nickel	Calx of nickel	Calx of nickel	Calx of nickel	Calx of nickel	Calx of nickel	Calx of nickel	Calx of nickel	Calx of nickel	Calx of nickel
14	Calx of lead	Calx of lead	Calx of lead	Calx of lead	Calx of lead	Calx of lead	Calx of lead	Calx of tin	Calx of lead	Calx of lead	Calx of lead	Calx of lead	Calx of lead	Calx of lead	Calx of lead	Calx of lead	Calx of lead	Calx of lead	Calx of lead	Calx of lead
15	Calx of tin	Calx of tin	Calx of tin	Calx of tin	Calx of tin	Calx of tin	Calx of tin	Calx of tin	Calx of tin	Calx of tin	Calx of tin	Calx of tin	Calx of tin	Calx of tin	Calx of tin	Calx of tin	Calx of tin	Calx of tin	Calx of tin	Calx of tin
16	Calx of copper	Calx of copper	Calx of copper	Calx of copper	Calx of copper	Calx of copper	Calx of copper	Calx of copper	Calx of copper	Calx of copper	Calx of copper	Calx of copper	Calx of copper	Calx of copper	Calx of copper	Calx of copper	Calx of copper	Calx of copper	Calx of copper	Calx of copper
17	Calx of bismuth	Calx of bismuth	Calx of bismuth	Calx of bismuth	Calx of bismuth	Calx of bismuth	Calx of bismuth	Calx of bismuth	Calx of bismuth	Calx of bismuth	Calx of bismuth	Calx of bismuth	Calx of bismuth	Calx of bismuth	Calx of bismuth	Calx of bismuth	Calx of bismuth	Calx of bismuth	Calx of bismuth	Calx of bismuth
18	Calx of antimony	Calx of antimony	Calx of antimony	Calx of antimony	Calx of antimony	Calx of antimony	Calx of antimony	Calx of antimony	Calx of antimony	Calx of antimony	Calx of antimony	Calx of antimony	Calx of antimony	Calx of antimony	Calx of antimony	Calx of antimony	Calx of antimony	Calx of antimony	Calx of antimony	Calx of antimony
19	Calx of arsenic	Calx of arsenic	Calx of arsenic	Calx of arsenic	Calx of arsenic	Calx of arsenic	Calx of arsenic	Calx of arsenic	Calx of arsenic	Calx of arsenic	Calx of arsenic	Calx of arsenic	Calx of arsenic	Calx of arsenic	Calx of arsenic	Calx of arsenic	Calx of arsenic	Calx of arsenic	Calx of arsenic	Calx of arsenic
20	Calx of mercury	Calx of mercury	Calx of mercury	Calx of mercury	Calx of mercury	Calx of mercury	Calx of mercury	Calx of mercury	Calx of mercury	Calx of mercury	Calx of mercury	Calx of mercury	Calx of mercury	Calx of mercury	Calx of mercury	Calx of mercury	Calx of mercury	Calx of mercury	Calx of mercury	Calx of mercury
21	Calx of silver	Calx of silver	Calx of silver	Calx of silver	Calx of silver	Calx of silver	Calx of silver	Calx of silver	Calx of silver	Calx of silver	Calx of silver	Calx of silver	Calx of silver	Calx of silver	Calx of silver	Calx of silver	Calx of silver	Calx of silver	Calx of silver	Calx of silver
22	Calx of gold	Calx of gold	Calx of gold	Calx of gold	Calx of gold	Calx of gold	Calx of gold	Calx of gold	Calx of gold	Calx of gold	Calx of gold	Calx of gold	Calx of gold	Calx of gold	Calx of gold	Calx of gold	Calx of gold	Calx of gold	Calx of gold	Calx of gold
23	Calx of platina	Calx of platina	Calx of platina	Calx of platina	Calx of platina	Calx of platina	Calx of platina	Calx of platina	Calx of platina	Calx of platina	Calx of platina	Calx of platina	Calx of platina	Calx of platina	Calx of platina	Calx of platina	Calx of platina	Calx of platina	Calx of platina	Calx of platina
24	Water	Water	Water	Water	Water	Water	Water	Pure siliceous earth	Spirit of wine	Spirit of wine	Spirit of wine	Spirit of wine	Spirit of wine	Water	Spirit of wine	Water	Water	Water	Water	Water
25	Spirit of wine	Spirit of wine	Spirit of wine	Spirit of wine	Spirit of wine	Spirit of wine	Spirit of wine		Phlogiston	Phlogiston	Phlogiston	Phlogiston	Phlogiston	Spirit of wine	Phlogiston	Spirit of wine	Spirit of wine	Spirit of wine	Spirit of wine	Spirit of wine
26	Phlogiston	Phlogiston	Phlogiston	Phlogiston	Phlogiston	Phlogiston	Phlogiston							Phlogiston		Phlogiston	Phlogiston	Phlogiston	Phlogiston	Phlogiston

In the Dry Way.

	1 VITRIOLIC ACID	2 PHLOGISTICATED VITRIOLIC ACID	3 NITROUS ACID	4 PHLOGISTICATED NITROUS ACID	5 MARINE ACID	6 DEPHLOGISTICATED MARINE ACID	7 AQUA REGIA	8 ACID OF FLUOR	9 ACID OF ARSENIC	...
31	Phlogiston	Phlogiston	Phlogiston	Phlogiston	Phlogiston	Phlogiston	Phlogiston		Phlogiston	
32	Pure vegetable alkali						Pure ponderous earth	Lime	Lime	
33	Pure fossil alkali						Pure vegetable alkali	Pure ponderous earth	Pure ponderous earth	
34	Pure pond. earth						Pure fossil alkali	Pure magnesia	Pure magnesia	
35	Lime						Lime	Pure vegetable alkali	Pure vegetable alkali	
36	Pure magnesia						Pure magnesia	Pure fossil alkali	Pure fossil alkali	
37	Metallic calces						Metallic calces	Metallic calces	Metallic calces	
38	Pure volatile alkali						Pure volatile alkali	Pure volatile alkali	Pure volatile alkali	
39	Pure clay						Pure clay	Pure clay	Pure clay	

SINGLE ELECTIVE ATTRACTIONS.

In the Moist Way.

21 ACIDS or FAT	22 ACIDS or PHOSPHORUS	23 ACETOUS PHLOGISTICATED	24 ACID of PHLOGISTICATED BLOOD	25 ANIMAL ACID	26 PURE VEGETABLE ALKALI	27 PURE FOSSIL ALKALI MALI	28 PURE VOLATILE ALKALI	29 PURE PONDEROUS EARTH	30 LIME	31 PURE MAGNESIA	32 PURE CLAY	33 PURE SILICEOUS EARTH	34 WATER	35 VITAL AIR	36 PHLOGISTON	37 MATTER of HEAT SULPHUR	38	39 SALINE LIVER, or ALCOHOL SULPHUR	40	
1 Lime	Lime	Lime	Pure vegetable alkali	Pure ponderous earth	Vitriolic acid	Vitriolic acid	Vitriolic acid	Acid of sugar	Acid of sugar	Acid of sugar	Vitriolic acid	Acid of fluor	Pure vegetable alkali	Phlogiston	Nitrous acid	Vital air	Cale of lead	Cale of gold	Water	a
2 Pure ponderous earth	Pure ponderous earth	Pure ponderous earth	Pure fossil alkali	Pure fossil alkali	Nitrous acid	Nitrous acid	Nitrous acid	Acid of fluor	Acid of fluor	Acid of fluor	Nitrous acid	Acid of phlogisticated	Pure fossil alkali		Vitriolic acid	Æther	Cale of tin	Cale of silver	Æther	3
3 Pure magnesia	Pure magnesia	Pure magnesia	Pure volatile alkali	Pure vegetable alkali	Marine acid	Marine acid	Marine acid	Acidum perlatum	Acidum perlatum	Acidum perlatum	Marine acid	Marine acid			Dephlogisticated marine acid	Alcohol	Cale of silver	Cale of mercury	Essential oils	4
4 Pure vegetable alkali	Pure vegetable alkali	Pure vegetable alkali	Lime	Pure volatile alkali	Acid of fat	Acid of fat	Acid of fat	Vitriolic acid	Vitriolic acid	Vitriolic acid	Acid of sugar	Acid of sugar	Alcohol		Acid of arsenic	Pure volatile alkali	Cale of mercury	Cale of arsenic	Pure volatile alkali	5
5 Pure fossil alkali	Pure fossil alkali	Pure fossil alkali	Pure magnesia	Phlone acid	Fluor acid	Fluor acid	Fluor acid	Phlone acid	Phlone acid	Phlone acid	Acid of sugar	Acid of sugar	Arsenical volatile		Phlogistic acid	Water	Cale of antimony	Cale of antimony	Saline liquor	6
6 Pure volatile alkali	Pure volatile alkali	Pure volatile alkali	Pure magnesia	Phlogistic acid	Phlogistic acid	Phlogistic acid	Phlogistic acid	Acidum perlatum	Acid of fat	Acid of fat	Acid of sugar of milk	Acid of phlogisticated			Phlogistic acid	Editorial oils	Cale of bismuth	Cale of bismuth	Sulphur	7
7 Pure clay	Pure clay	Pure clay		Acid of sugar	Acid of sugar	Acid of sugar	Acid of sugar	Acid of sugar of milk	Acid of sugar of milk	Acid of sugar of milk	Acidum perlatum	Phlone acid	Vitriolated fossil alkali		Cale of platina	Glauk	Pure fixed vegetable alkali	Cale of copper		8
8 Cale of zinc	Cale of zinc	Cale of zinc	Pure clay	Acid of tartar	Acid of tartar	Acid of tartar	Acid of tartar	Nitrous acid	Acid of tartar	Acid of tartar	Acid of fat	Acid of fat	Æther		Cale of gold	Mercury	Pure volatile alkali	Cale of lead		9
9 Cale of iron	Cale of iron	Cale of iron	Cale of zinc	Acid of arsenic	Acid of arsenic	Acid of arsenic	Acid of arsenic	Marine acid	Acid of amber	Acid of amber	Acid of amber				Cale of silver		Pure ponderous earth	Cale of nickle		10
10 Cale of manganese	Cale of manganese	Cale of manganese	Cale of iron	Acid of amber	Acid of amber	Acid of amber	Acid of amber		Marine acid	Marine acid	Acid of sugar of milk				Cale of mercury		Lime	Cale of cobalt		11
11 Cale of cobalt	Cale of cobalt	Cale of cobalt	Cale of manganese	Acid of lemon	Acid of lemon	Acid of lemon	Acid of lemon	Acid of fat	Acid of fat	Marine acid					Cale of arsenic					12
12 Cale of nickle	Cale of nickle	Cale of nickle	Cale of cobalt	Acid of tartar	Acid of tartar	Acid of tartar	Acid of tartar	Phlone acid	Acid of ferret	Acid of ferret		Vitriolic acid			Cale of antimony	Pure magnesia	Cale of manganese			13
13 Cale of lead	Cale of lead	Cale of lead	Cale of nickle	Acid of ants	Acid of ants	Acid of ants	Acid of sats	Acid of arsenic	Acid of tartar	Acid of tartar		Acid of phlogisticated			Cale of bismuth		Cale of iron			14
14 Cale of tin	Cale of tin	Cale of tin	Cale of lead	Acid of milk	Acid of milk	Acid of milk	Acid of milk	Acid of arsenic	Acid of sats	Acid of sats		Acidum perlatum	Vitriolic acid		Cale of copper					15
15 Cale of copper	Cale of copper	Cale of copper	Cale of tin	Acid of benzoin	Acid of benzoin	Acid of benzoin	Acid of benzoin	Acid of sats	Acid of sats	Acid of sats		Acid of sats	Vitriolated vegetable alkali		Cale of tin					16
16 Cale of bismuth	Cale of bismuth	Cale of bismuth	Cale of copper	Aceous acid	Acidum perlatum	Aceous acid	Acid of milk	Acid of milk	Acid of milk	Acid of milk		Acid of milk	Vitriolated iron		Cale of lead					17
17 Cale of antimony	Cale of antimony	Cale of antimony	Cale of bismuth	Acidum perlatum	Acidum perlatum	Acidum perlatum	Acid of benzoin	Acid of benzoin	Acid of benzoin	Acid of benzoin		Acid of benzoin	Vitriolated iron		Cale of nickle					18
18 Cale of arsenic	Cale of arsenic	Cale of arsenic	Cale of antimony	Acid of sugar of milk	Acid of sugar of milk	Acid of sugar of milk	Acetous acid	Acetous acid	Acetous acid	Acetous acid		Acetous acid	Corrosive sublimate mercury		Cale of cobalt					19
19 Cale of mercury	Cale of mercury	Cale of mercury	Cale of arsenic	Acid of borax	Acid of borax	Acid of borax	Acid of borax	Acid of borax	Acid of borax	Acid of borax		Acid of borax			Cale of manganese					20
20 Cale of silver	Cale of silver	Cale of silver	Cale of mercury	Phlogisticated nitrolic acid	Phlogisticated nitrolic acid	Phlogisticated nitrolic acid	Phlogisticated nitrolic acid	Phlogisticated nitrolic acid	Phlogisticated nitrolic acid	Phlogisticated nitrolic acid		Phlogisticated nitrolic acid			Cale of iron					21
21 Cale of gold	Cale of gold	Cale of gold	Cale of silver	Phlogisticated nitrous acid	Phlogisticated nitrous acid	Phlogisticated nitrous acid	Phlogisticated nitrous acid	Phlogisticated nitrous acid	Phlogisticated nitrous acid	Phlogisticated nitrous acid		Phlogisticated nitrous acid			Cale of zinc					22
22 Cale of platina	Cale of platina	Cale of platina	Cale of gold	Aerial acid	Aerial acid	Aerial acid	Aerial acid	Aerial acid	Aerial acid	Aerial acid		Aerial acid								23
23 Water	Water	Water	Cale of platina	Acid of Prussian blue	Acid of Prussian blue	Acid of Prussian blue	Acid of Prussian blue	Acid of Prussian blue	Acid of Prussian blue	Acid of Prussian blue		Acid of Prussian blue								24
24 Spirit of wine	Spirit of wine	Water	Water	Water	Water	Water	Water	Water	Water	Water					Water				Alcohol	25
25 Phlogiston	Phlogiston	Spirit of wine		U.ctuous oils	Unctuous oils	Uctuous oils	Unctuous ash	Unctuous oils	Unctuous oils										Water	26
26				Sulphur	Sulphur	Sulphur	Sulphur	Sulphur	Sulphur											27
27				Metallic calces		Phlogiston	Phlogiston													28
28																				29

In the Dry Way.

Lime	Lime	Phlogiston	Water	Phlogisticated acid Acidum perlatum Acid of borax	Phlogisticated acid Acidum perlatum Acid of borax	Phlogisticated acid Acidum perlatum Acid of borax	Phlogisticated acid Acidum perlatum Acid of borax	Phlogisticated acid Acidum perlatum Acid of borax	Phlogisticated acid Acidum perlatum Acid of borax	Phlogisticated acid Acidum perlatum Acid of borax		Pure fixed alkali			Cale of platina	Pure fixed alkali	Manganese	Alcohol
Pure ponderous earth	Pure ponderous earth			Acid of arsenic Vitriolic acid	Acid of arsenic Vitriolic acid	Acid of arsenic Vitriolic acid	Acid of arsenic Vitriolic acid	Acid of fluor Vitriolic acid	Acid of arsenic Vitriolic acid	Acid of fluor Vitriolic acid		Sulphur	Phlogistic acid		Cale of gold	Iron	Water
Pure magnesia	Pure magnesia			Nitrous acid	Nitrous acid	Vitriolic acid	Vitriolic acid	Acid of amber	Acid of amber	Acid of fluor		Cale of lead			Cale of silver	Copper
Pure vegetable alkali	Pure vegetable alkali			Marine acid	Marine acid			Fluor acid	Fluor acid	Acid of fat					Cale of antimony	Tin
Metallic calces	Pure fossil alkali			Acid of fat	Acid of fat			Acid of fat	Acid of fat	Nitrous acid					Cale of bismuth	Lead
Pure clay	Pure volatile alkali			Acid of amber	Acid of amber			Acid of amber	Acid of amber	Marine acid					Cale of tin	
	Pure clay			Acid of milk	Acid of milk			Acid of amber	Acid of amber						Cale of nickle	Silver
				Acid of borax	Acid of borax			Acid of borax	Acid of borax	Acetous acid					Cale of cobalt	Cobalt
				Acetous acid	Acetous acid			Acetous acid	Acetous acid	Aerial acid					Cale of lead	Nickle
				Pure magnesia	Pure magnesia										Cale of bismuth	Bismuth
				Lime	Lime										Cale of mercury	Antimony
				Pure ponderous	Pure ponderous										Cale of manganese	Mercury
				Pure siliceous earth	Pure siliceous earth										Cale of iron	Arsenic
				Sulphur	Sulphur											

SINGLE ELECTIVE ATTRACTIONS.

In the Moist Way.

	41 ÆTHER	42 ESSENTIAL OILS	43 EXPRESSED OILS	44 CALX OF GOLD	45 CALX OF PLATINA	46 CALX OF SILVER	47 CALX OF MERCURY	48 CALX OF LEAD	49 CALX OF COPPER	50 CALX OF IRON	51 CALX OF TIN	52 CALX OF BISMUTH	53 CALX OF NICKEL	54 CALX OF ARSENIC	55 CALX OF COBALT	56 CALX OF ZINC	57 CALX OF ANTIMONY	58 CALX OF MANGANESE	59 CALX OF SIDERITE
1	ÆEther	ÆEther	ÆEther	Marine acid	ÆEther	Marine acid	Acid of fat	Vitriolic acid	Acid of sugar	Acid of sugar	Acid of fat	Acid of sugar	Acid of sugar	Marine acid	Acid of sugar	Acid of sugar	Acid of fat	Acid of sugar	Vitriolic acid
2	Alcohol	Alcohol	Alcohol	Aqua regia	Marine acid	Acid of tartar	Marine acid	Acid of fat	Acid of tartar	Acid of tartar	Acid of tartar	Acid of arsenic	Acid of sorrel	Acid of sugar	Acid of sugar	Acid of sorrel	Marine acid	Acid of sorrel	Nitrous acid
3	Essential oils	Pure fixed alkali	Pure fixed alkali	Nitrous acid	Aqua regia	Marine acid	Acid of sugar	Acid of sugar of milk	Marine acid	Vitriolic acid	Marine acid	Marine acid	Marine acid	Vitriolic acid	Marine acid	Marine acid	Acid of sugar	Acid of lemon	Marine acid
4	Expressed oils	Pure volatile alkali	Pure volatile alkali	Vitriolic acid	Nitrous acid	Vitriolic acid	Acid of amber	Acid of sorrel	Vitriolic acid	Acid of sugar of milk	Vitriolic acid	Acid of tartar	Vitriolic acid	Nitrous acid	Vitriolic acid	Acid of sugar of milk	Vitriolic acid	Phosphoric acid	Acid of sugar of milk
5	Water	Sulphur	Sulphur	Arsenical acid	Vitriolic acid	Acid of sugar of milk	Arsenical acid	Arsenical acid	Nitrous acid	Marine acid	Acid of sugar	Phosphoric acid	Acid of tartar	Acid of tartar	Acid of tartar	Nitrous acid	Nitrous acid	Acid of tartar	Acid of tartar
6	Sulphur			Fluor acid	Arsenical acid	Phosphoric acid	Phosphoric acid	Acid of fat	Acid of fat	Nitrous acid	Arsenical acid	Acid of fat	Acid of fat	Acid of sorrel	Acid of fat	Acid of fat	Acid of tartar	Marine acid	Acid of sorrel
7				Acid of tartar	Fluor acid	Nitrous acid	Vitriolic acid	Phosphoric acid	Arsenical acid	Acid of fat	Phosphoric acid	Acid of fat	Acid of fat	Phosphoric acid	Acid of sorrel	Acid of sorrel	Acid of sorrel		
8				Phosphoric acid	Acid of tartar	Vitriolic acid	Vitriolic acid	Phosphoric acid	Acid of fat	Acid of fat	Fluor acid	Acid of fat	Acid of tartar	Acid of sorrel	Acid of fat	Acid of sorrel	Acid of sorrel	Marine acid	Vitriolic acid
9					Phosphoric acid	Arsenical acid	Acid of sugar of milk	Acid of sorrel	Arsenical acid	Phosphoric acid	Nitrous acid	Marine acid	Phosphoric acid	Acid of sugar of milk	Phosphoric acid	Acid of tartar	Acid of sugar of milk	Vitriolic acid	Vitriolic acid
10 / 11				Phosphoric acid / Acid of fat	Phosphoric acid / Acid of fat	Fluor acid / Acid of tartar	Acid of tartar / Acid of lemon	Marine acid / Acid of lemon	Phosphoric acid / Acid of forrel	Arsenical acid / Acid of forrel	Acid of amber / Acid of forrel	Nitrous acid / Fluor acid	Fluor acid / Acid of amber	Fluor acid / Acid of amber	Fluor acid / Acid of amber	Phosphoric acid / Acid of amber	Phosphoric acid / Acid of amber	Nitrous acid / Acid of fugar of milk	
12				Acid of Prussian blue	Acid of forrel	Acid of forrel	Nitrous acid	Fluor acid	Acid of amber	Fluor acid	Fluor acid	Acid of fugar of milk	Acid of amber	Acid of amber	Acid of amber	Acid of amber	Fluor acid	Acid of fat	
13					Acid of lemon	Acid of lemon	Nitrous acid	Acid of lemon	Acid of amber	Acid of amber	Acid of fugar of milk	Acid of amber	Acid of lemon	Acid of lemon	Acid of lemon	Fluor acid	Fluor acid	Artificial acid	
14 / 15 / 16				Acid of lemon / Acid of ants / Acid of tartar	Acid of lemon / Acid of ants / Acid of milk	Acid of ants / Acid of ants / Acetous acid	Acetous acid / Acid of ants / Acid of Prussian blue	Acid of ants / Acid of milk / Acetous acid	Acid of lemon / Acid of milk / Acid of milk	Acid of lemon / Acid of milk / Acid of milk	Acid of lemon / Acid of ants / Acid of milk	Acid of lemon / Acid of ants / Acid of milk	Acid of ants / Acid of milk / Acetous acid	Acid of ants / Acid of milk / Arsenical acid	Acid of ants / Acid of milk / Acetous acid	Acid of amber / Acid of milk / Acid of milk	Arsenical acid / Acid of ants / Acid of milk	Artificial acid / Acid of ants / Acid of milk	
17 / 18				Acid of Prussian blue	Acetous acid / Acid of amber	Acid of amber / Acid of Prussian blue	Aerial acid	Acid of borax / Acid of Prussian blue	Acetous acid / Acid of borax	Acetous acid / Acid of borax	Acetous acid / Acid of borax	Acetous acid	Arsenical acid / Acid of borax	Acetous acid	Arsenical acid / Acid of borax	Acetous acid / Acid of borax	Acetous acid / Acid of borax	Acetous acid	
19				Acid of Prussian blue		Aerial acid		Aerial acid	Acid of Prussian blue	Acid of Prussian blue	Acid of Prussian blue	Acid of Prussian blue	Acid of Prussian blue	Acid of Prussian blue	Acid of Prussian blue	Acid of Prussian blue	Acid of Prussian blue	Acid of Prussian blue	Acid of Prussian blue
				Fixed alkali / Volatile alkali		Pure volatile alkali	Aerial acid	Pure fixed alkali	Aerial acid / Pure volatile alkali / Expressed oils	Aerial acid	Fixed alkali / Volatile alkali	Aerial acid / Volatile alkali	Aerial acid / Volatile alkali	Volatile alkali / Unctuous oils	Aerial acid / Volatile alkali	Aerial acid / Volatile alkali	Aerial acid		Fixed alkali / Volatile alkali
																			Water

In the Dry Way.

	GOLD	PLATINA	SILVER	MERCURY	LEAD	COPPER	IRON	TIN	BISMUTH	NICKEL	ARSENIC	COBALT	ZINC	ANTIMONY	MANGANESE	SIDERITE
	Mercury	Arsenic	Lead	Gold	Gold	Gold	Nickle	Zinc	Lead	Iron	Nickle	Iron	Copper	Iron	Copper	Cobalt
	Copper	Gold	Copper	Silver	Silver	Silver	Cobalt	Copper	Silver	Cobalt	Cobalt	Nickle	Antimony	Copper	Iron	Nickle
	Silver	Copper	Mercury	Platina	Copper	Arfenic	Manganefe	Antimony	Gold	Arfenic	Copper	Arfenic	Tin	Tin	Gold	Iron
	Lead	Tin	Bismuth	Lead	Mercury	Iron	Arfenic	Gold	Mercury	Copper	Iron	Copper	Mercury	Lead	Silver	Manganefe
	Bismuth	Bismuth	Tin	Tin	Bismuth	Mangauefe	Copper	Silver	Antimony	Gold	Tin	Gold	Silver	Silver	Tin	
	Tin	Zinc	Gold	Zinc	Tin	Zinc	Gold	Lead	Tin	Tin	Silver	Platina	Gold	Bismuth		
	Antimony	Antimony	Antimony	Bismuth	Antimony	Antimony	Silver	Iron	Copper	Antimony	Lead	Tin	Cobalt	Zinc		
	Iron	Nickle	Iron	Copper	Platina	Platina	Tin	Manganefe	Platina	Platina	Gold	Antimony	Arfenic	Gold		
	Platina	Cobalt	Manganefe	Antimony	Arfenic	Tin	Antimony	Nickle	Nickle	Bifmuth	Platina	Zinc	Platina	Platina		
	Zinc	Manganefe	Zinc	Arfenic	Zinc	Lead	Platina	Arfenic	Iron	Lead	Zinc		Bifmuth	Mercury / Arfenic / Cobalt		
	Nickle	Mercury	Arfenic	Iron	Nickle	Nickle	Bifmuth	Platina	Zinc	Silver	Antimony		Lead / Nickle / Iron			
	Arfenic		Nickle		Iron	Bifmuth	Lead	Bifmuth		Zinc						
	Cobalt		Platina			Cobalt	Mercury	Cobalt								
	Manganefe					Mercury										
	Saline liver of fulphur	Saline liver of fulphur	Saline liver of fulphur	Saline liver of fulphur	Saline liver of fulphur	Saline liver of fulphur	Saline liver of fulphur	Saline liver of fulphur	Saline liver of fulphur	Saline liver of fulphur	Saline liver of fulphur	Saline liver of fulphur	Saline liver of fulphur	Saline liver of fulphur	Saline liver of fulphur	Saline liver of fulphur
	Sulphur															

www.ingramcontent.com/pod-product-compliance
Ingram Content Group UK Ltd.
Pitfield, Milton Keynes, MK11 3LW, UK
UKHW020410010325
455677UK00029B/831